CANADA'S FORESTS

Forest History Society Issues Series

The Forest History Society was founded in 1946. Since that time, the Society, through its research, reference, and publications programs, has advanced forest and conservation history scholarship. At the same time, it has translated that scholarship into formats useful for people with policy and management responsibilities. For seven decades the Society has worked to demonstrate history's significant utility.

The Forest History Society's Issues Series is one of the Society's most explicit contributions to history's utility. The Society selects issues of importance today that also have significant historical dimensions. Then we invite authors of demonstrated knowledge to examine an issue and synthesize its substantial literature, while keeping the general reader in mind.

The final and most important step is making these authoritative overviews available. Toward that end, each book is distributed to people with management, policy, or legislative responsibilities who will bene-fit from a deepened understanding of how a particular issue began and evolved.

The Issues Series – like its Forest History Society sponsor – is non-advocatory and aims to present a balanced rendition of often con-tentious issues. The pages that follow document the importance of Canadian forests not only to Canada but to the world. The author traces the changes Canadian forests have experienced over 10,000 years and describes the interaction of humans with Canadian forests during that time. In contrast to many other parts of the world, and due in part to the forest's resiliency and in part to policies developed during the last 150 years, Canadian forests remain healthy and largely intact.

The Society gratefully acknowledges financial support from Natural Resources Canada, the Weyerhaeuser Company Foundation, the Henry P. Kendall Foundation,and J.D. Irving, Limited, for this sixth title in the Issues Series.

Other Issues Series titles available from the Forest History Society:

American Forests: A History of Resiliency and Recovery
Newsprint: Canadian Supply and American Demand
Forest Pharmacy: Medicinal Plants in American Forests
American Fires: Management on Wildlands and Forests
Forest Sustainability: The History, the Challenge, the Promise

Canada's Forests

A History

Ken Drushka

Forest History Society Issues Series
Forest History Society

McGill-Queen's University Press
Montreal & Kingston • London • Ithaca

The Forest History Society is a nonprofit educational institution dedicated to the advancement of historical understanding of human interaction with the forest environment. The society was established in 1946. Interpretations and conclusions in FHS publications are those of the authors; the society takes responsibility for the selection of topics, the competence of the authors, and their freedom of inquiry.

Forest History Society
701 Vickers Avenue
Durham, North Carolina 17701
(919) 682-9319

ISBN 0-7735-2660-9 (cloth)
ISBN 0-7735-2661-7 (paper)
Legal deposit third quarter 2003
Bibliothéque nationale du Québec

Printed in Canada on acid-free paper that is 100% ancient forest free (100% post-consumer recycled), processed chlorine free.

This book has been published with the help of a grant from the Forest History Society.

McGill-Queen's University Press acknowledges the financial support of the Government of Canada through the Book Publishing Industry Development Program (BPIDP) for its activities. It also acknowledges the support of the Canada Council for the Arts for its publishing program.

National Library of Canada Cataloguing in Publication

Drushka, Ken
    Canada's forests : a history / Ken Drushka.

(Forest History Society issues series)
Includes bibliographical references and index.
ISBN 0-7735-2660-9

    1. Sustainable forestry–Canada–History. 2. Forest management–Canada–History. 3. Forests and forestry–Canada–History. I. Forest History Society II. Title. III. Series.
SD567.D78 2003   333.75'0971   C2003-903362-7

www.mqup.ca

CONTENTS

Figures

OVERVIEW

Canadian forests represent about 10 percent of the world's forest. After 10,000 years of human use, Canadian forests still exist virtually intact. Since European colonization, about 26 million hectares or only 6 percent of Canada's forestland has been converted to other uses. There is more forestland in Canada today than there was 75 years ago.

- Today's forests in Canada began developing after the last period of glaciation, about 10,000 years ago. The forests have advanced and retreated and changed composition during the cyclical periods of global warming and cooling.

- Canada's aboriginal population manipulated the forest and the landscape to meet their needs. Some permanent settlements, agriculture, and the extensive use of fire marked their impact on the forest.

- The fur trade was Canada's first forest harvest. As the dominant economic activity for 200 years in North America, it provided the incentive and financial support for European exploration and settlement. Clearing for home sites and agriculture was the main cause for loss of forestland during the colonial period.

- The industrialization of Canada made great demands on Canadian forests for constructions of railways and bridges. Fires started by wood and coal burning steam engines had extensive effect on the landscape. In 1900, the state of Canada's forests was at its lowest point since the period of glaciation.

- The conservation movement in the early 1900s was marked by the opening of the first forestry schools in Canada, the establishment of parks, reforestation efforts, protection from fires, and the application of scientific management.

- The outbreak of World War I and II interrupted conservation efforts, but the move toward sustained yield for Canada's forests moved forward, buoyed by the economic resurgence of the postwar period. Improvements in timber-processing technology, especially utilization of small diameter species, helped satisfy increasing demand.

- In the late 1900s, forest health concerns and challenges with the application of sustained-yield and multiple-use concepts led to a new focus on managing sustainable forests, articulated in Canada's National Forest Strategy adopted at the 1992 Forest Congress and refined in 1998 and 2003.

- Challenges to forest management include forest fragmentation, fire policy, amount of area in preserved status, and levels of investment in forest improvement activities.

- Today, Canada retains a larger portion of its original forest than any other nation on earth.

CANADA'S FORESTS

INTRODUCTION

Canada's forests are the country's dominant geographical feature. They carpet the landscape in a vast, largely unbroken swath, covering three-quarters of the country below the northern tree line.

These forests define the country. They shape its character and its image. They are a determining factor for most life forms – including humans – that inhabit the northern portion of the North American continent.

Since long before the first European settlers arrived, the forests have dominated the consciousness of the people who inhabit this land. Forests shape the economic, social, and cultural life of Canadians, and their condition and use is critical to the welfare of the nation.

Despite their importance, the forests and their disposition are barely understood by the large majority of citizens. Yet most Canadians manifest a passionate concern for their forests. This has not always been the case, but by the dawn of the twenty-first century, Canadians had realized that without its forests, the country could not exist.

Today, people in other parts of the world also appreciate the importance of Canada's forests. Given their global importance – the role the forests play in regulating the basic necessities of life on the planet and the human needs they meet – these feelings are not misplaced. Canada's forests account for 10 per cent of all the forests in the world. They help maintain the oxygen required by life on the planet. They provide a substantial portion of the world's fresh water supply and habitat for the country's diverse non-domesticated organisms. Canadian forest-product industries, in addition to meeting domestic needs, are the world's largest exporters of timber-based goods. Currently, Canada supplies

50 per cent of global lumber exports and 56 per cent of newsprint exports. The entire world has a stake in Canada's forests.

Over time the condition of Canada's forests has changed. Human use, and at times misuse, of these forests has varied. In some cases the result has been the conversion of forests to other uses. In a few instances it has meant the destruction of forests. In still others human impact has been short lived. Centuries of use have provided a large and growing body of knowledge about how to manage and care for the country's forests.

During the late twentieth century an often-fierce debate raged over the use of forests in some parts of Canada. Often this debate has been healthy and led to a greater appreciation and under-standing of forests. Sometimes it has raised confusion and con-cern about the current condition of the country's forests. Through a historical comparison, this book provides a framework to assist in assessing the current state of Canada's forests.

CHAPTER 1

THE CANADIAN FOREST

Origins of the Forest

Canada's forests are very young. In geological time, they have existed for only a brief moment, since the last glaciation ended about ten thousand years ago. At several periods during the last two to three million years, most of Canada was covered with ice. Each "Ice Age" lasted for about one hundred thousand years. Between these epochs the land was inhabited by various forms of plant and animal life, including forests of one sort or another.

The interglacial forests were likely much different from those that exist today. In the high Arctic, for instance, traces of ancient forests can be found, suggesting the previous existence of a forest much larger in extent than that which is now present. Elsewhere, near Smithers in central British Columbia, for example, fossil remains have been found of metasequoia (dawn redwood) and gingko trees, species that now grow wild only in remote areas of China.

During the last period of glaciation about ten thousand years ago, all of Canada – except for a portion of northwestern Yukon and small, isolated refuges in the Cypress Hills, the Swan Hills, and parts of coastal B.C. – was covered with ice up to four thousand metres thick. The ice formed in two main sheets: the Cordilleran sheet, which covered present-day British Columbia, and the Laurentide sheet, which covered most of the rest of Canada and the northeastern United States. For part of this glacial period, a corridor along the eastern margins of the Rocky Mountains remained ice free.

As temperatures rose and the ice retreated, plants and animals began to re-populate the land. They migrated from the south, the

Figure 1.1 The last glacial period, 10,000 B.C.

refuges, and along the ice-free corridor from the north. It is widely believed the first human beings entered North America at this
time, crossing the Bering land bridge from Siberia and moving
south through the ice-free regions of northern Yukon and Alaska.
After populating the southern portion of the North American
continent, they migrated north as the ice sheets receded, other
animals and plants moving with them.

The forests that now exist have developed since that time. They
have advanced, retreated, and changed composition during cyclical intervals of global warming and cooling. And while they may

seem ancient and enduring to human beings, in geological history they are only a fleeting phenomenon.

When Europeans first arrived and began settling what is now Canada, the forests covered an estimated 444 million hectares out of a land mass of 922 million hectares. They comprised an estimated 140,000 species of plants, animals, and micro-organisms, including 180 species of trees, their most obvious component. Much of the country's unforested lands lay in the northern territories where contemporary climatic conditions were not conducive to forest growth. Today, almost twenty million hectares of alpine tundra and glaciers in the western mountains have no tree cover. The treeless portion of the central Canadian prairie is caused by a lack of moisture in combination with extremes in temperature, but this region is only a fraction of the size of the unforested area in the north.

The extent of the forests is a somewhat imprecise measurement. It is difficult to determine exactly where a forest begins or ends and where grassland or tundra takes over. Similarly, it is rare that precise boundaries are found between forest types, as there are usually transition zones where one type merges with another.

As our ability to measure and take inventory of the forests and their myriad components improves, our estimates of forest area change, making comparisons between different points in time difficult and unreliable. Even the most up-to-date estimates are based upon calculations made in different parts of the country over a period of two to three decades. As well, definitions of what constitutes a forest or a forest type change over time.

And finally, the diversity of Canadian forests makes it difficult to generalize about them. There are ten different major forest types or regions in the country, with wide variations in composition.

The Forest Regions

The character of the forests that occupied the glaciated landscape was determined by the interaction of soils, climate, and the species inhabiting adjacent lands. The present-day variation in these forests is a reflection of different geographical conditions in Canada. The rugged mountainous terrain of British Columbia, combined with the inflow of moist air from the Pacific Ocean, has produced diverse forest types within one

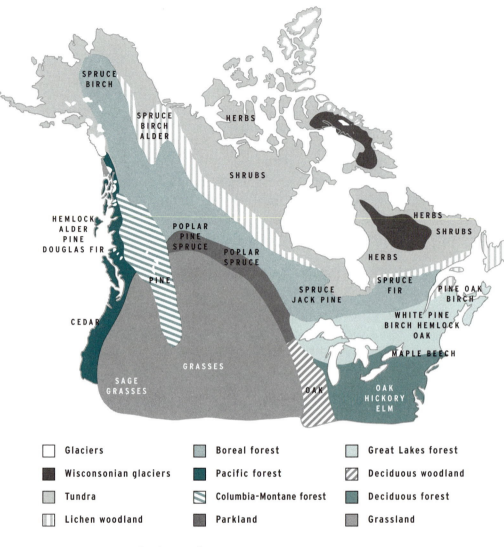

Figure 1.2 The forest advances, 5,000 B.C.

province. The central part of the country, stretching in a great arc between northwestern Yukon to Newfoundland, is covered with a fairly uniform type of forest. The southernmost part of the country around the Great Lakes is covered with a distinctly different forest, as are the prairie plains and most of the Maritime provinces. This diversity of forest types all within one nation is

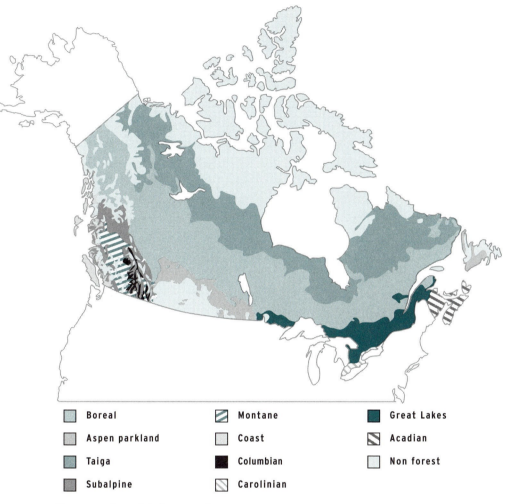

Boreal

Aspen parkland

Taiga

Subalpine

Montane

Coast

Columbian

Carolinian

Great Lakes

Acadian

Non forest

Figure 1.3 Canada's forest regions

unusual. While the Siberian forests may be larger in area, and tropical rainforests may contain more diverse species, few countries other than Canada have such a variety.

Boreal

By far the largest forest region in Canada, the boreal forest accounts for 76 per cent of the country's forested land area. It covers a wide range, from the delta of the Mackenzie River at 68 degrees latitude in the northwest, to 47 degrees latitude in southern

Quebec. It is an immense forest of relatively few tree species, with three sub-zones.

Taiga

The northernmost boreal forest, known as the taiga, comprises almost one-half of the boreal region, gradually emerging out of the Arctic tundra in the form of scattered dwarf trees struggling for survival in protected sites and along streams. The trees increase in size and density towards the south. The dominant species are white spruce and black spruce, with lesser amounts of tamarack.

Few large mammals but many small ones live in the taiga, along with large populations of insects and of migrating birds that nest in this region.

Central Boreal

The central boreal sub-zone is more heavily treed. The trees are larger and the species more diverse. Again, white spruce and black spruce are dominant, but there are also significant populations of balsam fir, jack pine, white birch, and trembling aspen.

Portions of this sub-zone occasionally push far into the northern boreal. A 480-kilometre-long stretch follows the Mackenzie River north from Great Slave Lake; another portion occupies the Churchill River and other watersheds in eastern Labrador; and a third area lies along the eastern shore of James Bay. In Newfoundland, northern and central boreal sub-zones are intermingled throughout the province, largely depending on variations in soil conditions.

Aspen Parkland

The southern boreal or aspen parkland sub-zone occupies a strip along the northern edge of the prairie grasslands, stretching from the Rocky Mountains in southern Alberta to southeastern Manitoba. In this forest, conifers gradually give way to stands of aspen intermixed with willow, which become smaller and less frequent further south in the grasslands. Isolated pockets of aspen are found throughout the grasslands and in one unique area of grassland deep in the central boreal – the Peace River country that straddles the northern Alberta-B.C. border.

On the whole the boreal forest is characterized by its climate, which tends to extremes. A short summer of long, hot days contrasts with long, cold winters during which plants are dormant. The northern boreal contains large areas of permanently frozen ground – permafrost – while large portions of the rest of the region consist of poorly drained muskeg bogs. About one-fifth of the region is made up of peat bogs.

The boreal is one of the world's great reservoirs of fresh water, found in thousands of streams, lakes, and wetlands. In part this is due to the relatively low levels of transpiration and evaporation of rain and snowfall back into the atmosphere in this region. Globally, almost 70 per cent of precipitation is returned to the atmosphere; in Canada as a whole the rate is only 40 per cent, the remainder entering lakes, streams, and underground water tables. This feature of Canadian hydrology is of immense ecological, economic, and social importance, providing extensive riparian habitats, hydroelectric potential, transportation opportunities, and other benefits. Most of the boreal watersheds are undeveloped or only lightly developed, providing a vast volume of fresh, clean water.

The central and southern portions of the boreal provide habitat for a wide variety of wildlife. Large mammals are plentiful, as are birds, fish, and insects. From a global perspective the Canadian boreal is one of the world's largest intact wildlife habitats.

Acadian

This region, encompassing Nova Scotia, Prince Edward Island, and the southern two-thirds of New Brunswick, is part of a forest type that extends south into the northeastern U.S. Trees in this region are a diverse mixture of coniferous species found in the boreal and adjacent Great Lakes–St Lawrence regions and deciduous species found south of the region. In some respects, and largely because it has no trees unique to itself, this is as much a transitional forest as a type of its own.

Today the primary conifers are red spruce, balsam fir, white spruce, black spruce, eastern hemlock, tamarack, white cedar, red pine, jack pine, and white pine. Deciduous species include sugar maple, red maple, silver maple, yellow birch, white birch, grey birch, beech, white, black and red ash, bur oak, red oak, basswood, hornbeam, trembling aspen, and large-tooth aspen.

Much of this area is subject to severe maritime climatic conditions, which inhibit tree growth and create areas of shrub-covered heath, particularly in Nova Scotia. The most heavily treed areas are in the fertile, protected valleys and interior New Brunswick.

The Acadian forest, perhaps more than any other in Canada, has suffered the devastating effects of wind and fire. It lies in the tail-end of the Atlantic hurricane zone, and on various occasions vast areas of timber have blown down. In 1798, for instance, more than a half-million hectares were destroyed in Nova Scotia by a hurricane. Fourteen years earlier, in 1784, two-thirds of Prince Edward Island had burned.

Great Lakes–St Lawrence

This is the second-largest forest region in Canada, stretching from eastern Gaspé and northern New Brunswick up the St Lawrence River and along the northern shore of the Great Lakes into eastern Manitoba. It bears many similarities to the Acadian forest, but a wider range of coniferous species, such as eastern red and white pines and eastern hemlock, gives it a character of its own. It is a true transition forest, composed of a blending of the boreal and the large deciduous forests of the northeastern United States. The major deciduous species are maple, oak, basswood, aspen, ash, elm, and birch.

A wide variation in geographical influences – numerous soil types and landscapes caused by the last glaciation and climatic conditions modified by the lakes – produces a diverse mixed-wood forest ecology within the region, with many different combinations of tree species. This ecological diversity is the primary characteristic of the region.

Carolinian

This small area of southwestern Ontario, sometimes called the Deciduous region, is the northernmost extension of the large central U.S. hardwood forest. As found by Europeans, it was a magnificent deciduous forest with a wide range of species including maple, oak, ash, and elm, many of them unique to Canada. Most of this forest has been eradicated for agriculture and other forms of development. Vestiges of the original forest are found only in small isolated stands, in woodlots, or in hedgerows. Similarly, its wildlife and native plant populations have been the most heavily reduced; about 60 per cent of Canada's endangered forest-dwelling species are found in this region.

A dozen or more deciduous species here grow nowhere else in Canada, including the commercially valuable black walnut. Many of the conifers found in the Great Lakes–St Lawrence forest are scattered throughout the region.

The ecological make-up of the Carolinian forest is largely determined by its mild climate, a result of its location between three Lakes: Ontario, Huron, and Erie. It is in the southernmost part of Canada, with a Mediterranean latitude. In combination with the highly fertile soils created by the rapid composition of organic matter, this makes it biologically an extremely productive area.

Columbia

West of the Rocky Mountains, throughout much of the interior of British Columbia, there is a complex intermingling of forest regions. This intricate mixing of forest types is caused by the interactions of terrain and climate. Within a single watershed, for example, two or more types of forest may exist, each at different altitudes. The Columbia region in southeastern B.C., sometimes known as the interior wet belt, provides the best example of this condition.

A mid-altitude region, it is restricted to those portions of mountain slopes above the valley bottoms and below the sub-alpine zones. These pockets of Columbia forest are found in the Columbia River valley, parts of the Kootenay River system, and westward into the Okanagan and Cariboo districts. The slopes where these forests grow are exposed to moist air blowing in from the Pacific, which precipitates as it rises to cross the province's north-south mountain ranges. This creates a damp inland forest similar in composition to forests in the coastal region.

Major coniferous species include Douglas fir, western hemlock, western red cedar, western white pine, and grand fir – all of which are found on the coast as well. Receiving less moisture and with a shorter growing season, these trees normally do not grow to the large sizes found on the coast but are generally much larger than conifers in other parts of the B.C. interior. One indigenous conifer, western larch, is not found on the coast but is prolific in this region. The dominant deciduous species is black cottonwood, with trembling aspen and white birch also present in significant numbers.

Montane

Comprised of low and mid-elevation forests, this is a relatively dry forest region lying in the rain shadows of the coastal range in the west and the Selkirk and Purcell ranges in the interior of British Columbia. It occupies an extensive area in the central interior around Prince George.

Trees are much smaller here than in coastal or Columbia forests. Stands tend to be dense in northern portions of the region and scattered in the south where grasslands are intermingled with forest. The prevalent species is lodgepole pine, with large populations of Douglas fir, Engelmann spruce, trembling aspen, and, in the south, ponderosa pine. Small pockets of montane forest are found east of the Rockies in the broad valleys at Waterton Lakes, Banff, and Jasper.

Subalpine

One of the most extensive forest regions in B.C., along with the montane and boreal, the subalpine region covers the upper levels of mountain regions throughout the interior and in small isolated patches on Vancouver Island and much of the province's north-central interior. It also extends along the eastern slopes of the Rockies in Alberta.

The determining condition in this region is climate. Like the boreal, it occupies the outer limits of land capable of supporting forest cover, the high mountains and the more northerly regions. Low winter temperatures, a short growing season, and cool nights restrict growth, producing relatively small trees.

The dominant tree species are Engelmann spruce and alpine fir, interspersed with populations of other species from the boreal, montane, and Columbia regions. Subalpine forests occupy the headwaters of most major river systems throughout B.C. and southern Alberta.

Coastal

The forest that occupies the middle and lower portions of coastal mountains, as well as the islands and mainland river valleys, is part of a forest stretching from central California to the top of the Alaska panhandle. One of the most biologically productive and economically valuable forests in the world, it is a dramatic and inspiring environment for humans.

Its character is determined by the existence of a coastal mountain range that precipitates large amounts of moisture from heavily laden air moving inland from the Pacific. The maritime influence provides relatively mild winters and a long growing season. The result has been the evolution of tree species that grow to very large sizes.

The dominant species is western hemlock. Douglas fir grows in dense stands in the southern portions but is absent from northern areas where Sitka spruce and western hemlock dominate. Western red cedar is abundant throughout the region on lower, wetter sites. Most other species are also conifers, with scattered individuals or small stands of deciduous species – red alder, bigleaf maple, and black cottonwood. Garry oak and arbutus are found only in the southern portion of the Canadian coastal region.

Like its Acadian cousin, the coastal forest is subject to extensive storm damage and, in the drier regions along eastern Vancouver Island, fire. Although it is relatively free of heavy insect attacks, some species are subject to major infestations of root disease.

In addition to its timber resources, the coastal forest is an important habitat for diverse fish and wildlife species. In particular, it provides spawning habitat for several species of salmon, which constitute one of the world's major commercial fisheries. These fish require freshwater streams in which to lay their eggs and, in some cases, rear young for the first year of their lives. After the deciduous region in Ontario, the coastal forest contains the second largest number of endangered species.

Distinct divisions between the different forest regions rarely occur. They blend and overlap in transition zones containing characteristics and species of each. Often pockets of one forest type are found in others, either due to local climatic or geologic conditions or as the result of human interference. Nor are the boundaries between these regions static. There is a constant shifting of the regions because of long-term changes in climate.

Most Canadian forests, except for those in the coastal regions, the high alpine areas, and the B.C. Interior wet belt, are less than 150 years old. They have been burned periodically and are subject to insect and disease attacks that limit their age.

CHAPTER 2

EARLY FOREST USE

The Prehistoric Forest

The nature of the forest that existed before the arrival of European explorers is a matter of debate. It is increasingly evident that Canada's aboriginal people modified the forests in which they lived, in some cases extensively. The long-held idea of a pristine forest, a blanket of green stretching from sea to sea – with a brief detour around the western plains – and unsullied by human intervention is not supported by historic or scientific evidence.

What is clear is that the forests that reclaimed the sterile, glaciated landscape as the ice receded evolved in tandem with the occupation of this country by human beings. It is also clear that the earliest inhabitants had an often-significant impact on the forests.

Aboriginal forest utilization and management took several forms. At the most basic level the people utilized parts of the forest for food, clothing, shelter, and fuel. They hunted wildlife for food and skins. They ate forest plants and used them for medicine. They felled and shaped trees into houses, boats, and other useful articles. They gathered wood to fuel their fires. A relatively small number of people were using their own muscle power to perform these tasks, so the impact on the forests for these consumptive uses was negligible. In almost all cases, utilization levels would have been much less than the reproduction and growth rates of the species involved.

Estimates of the aboriginal population of what is now Canada vary from 500,000 to two million people at the time of contact with Europeans. With one notable exception (the Huron in Ontario), they were primarily hunters and gatherers, moving across their territories with the change of seasons. They did not settle permanently in one location and thus did not leave a heavy footprint upon the land. Even the tribes on the Pacific coast, who made relatively intensive use of western red cedar and other tree species, had little or no enduring impact on the forests where they lived.

The most extensive aboriginal impact on forests was through the use, both accidental and deliberate, of fire. The first inhabitants of the Acadian forest in Nova Scotia and New Brunswick maintained and intensified the deciduous character of large areas of the forest with annual light burns of undergrowth and encroaching shade-tolerant coniferous species.

Throughout the boreal forest, aboriginals set fires to create meadows in which plants that attracted wildlife could flourish. They also undertook controlled spring burns of those parts of the forest containing trees killed by insects or blown down in order to prevent uncontrollable summer and fall fires started by lightning. In some areas, fires were set to drive game into traps for slaughter. Fire was also used as a weapon against other tribes and, occasionally, traders and settlers.

Figure 2.1 Indians firing prairie grasslands. Watercolour of native life by Alfred J. Miller, 1867. Source: National Archives of Canada C-000432

Fire was used extensively in the southern boreal forests and on the prairie grasslands. It is generally accepted that these fires, many of them deliberately set and controlled, substantially increased the deforested area of the plains, creating additional habitat for buffalo, deer, and other species important to local aboriginal economies.

Even on the moist Pacific coast, aboriginal people made extensive use of fire. From California north along the eastern coast of Vancouver Island and the adjacent mainland, early explorers described large unforested areas where camas bulbs and other plant foods grew. In the account of his 1882 trip through British Columbia, Newton Chittenden described the existence of large unforested meadows in various coastal locations including the Fraser, Skeena, and Nass valleys.

In Robert Brown's account of the 1864 Vancouver Island Exploring Expedition, one of his colleagues reported of the southern San Juan Valley: "The country was thickly wooded and mountainous but he saw some fine green prairies of 2 or 3000 acres."

The dense unburned forests of the Pacific northwest provided little food for humans or animals, while burned areas produced lush plant growth for a decade or more before trees once again claimed the site. These forest openings provided an abundance of berries and succulent plants that attracted deer and elk. This manipulation of the landscape was a transitional stage in the evolution from a hunting-gathering to an agricultural economy.

At the time of contact, this evolution was most advanced among the tribes inhabiting the deciduous forest region surrounding Lake Ontario and Lake Erie. Decaying litter from the hardwood species dominating these forests produced rich, fertile soil, and agricultural practices developed by aboriginal tribes to the south were adopted and refined before the first Europeans arrived in the area. Most of these people were Iroquois who lived in permanent settlements and cultivated fields of corn, beans, tobacco, sunflowers, and squash. Corn was the first agricultural crop introduced to the area in about 500 A.D.

In Canada the largest and most advanced group of aboriginal farmers were the Huron who occupied the tract of land between Georgian Bay and Lake Simcoe in southwestern Ontario. When Samuel de Champlain first visited this area in 1615, about thirty thousand Hurons occupied twenty-five villages, each surrounded by cultivated fields of perhaps ten thousand hectares in total.

Champlain's visit to Huronia was an early stage in a series of events of great consequence for Canada's forests. This change began with a shift in aboriginal forest use that constituted the first extensive commercial use of the North American forests: the fur industry.

The First Forest Harvest

The North American fur industry started in Newfoundland as an offshoot of the fishing industry, which began in the early 1500s. Large European fleets made annual expeditions to the inshore fishing grounds to catch cod. Shore installations were established to dry the fish before shipment back to Europe. These installations, with their docks, residential buildings, warehouses, drying racks, and other constructions, utilized large volumes of timber growing in the coastal regions of Newfoundland. This was the first significant use of American forests by Europeans, and the regulations adopted to control this usage constitute the first forest laws of the continent.

Shore workers who tended the fishing camps began trading European-made goods – axes, knives, iron kettles, cloth, blankets, and other items – to the resident Beothuk people for furs, especially beaver. Demand for this species was spurred by a fashion craze for wide-brimmed felt hats in Europe. Beaver pelts were valued for their fur, which produced high quality felt.

Demand for beaver and other furs increased dramatically, fostering a growth in trade on the continental mainland. The French, who controlled the St Lawrence and the Great Lakes, extended this trade along major rivers into the heart of the North American continent. They entered into complex trading agreements with various aboriginal tribes, each seeking to protect and expand their trading territory. Firearms, needed by the aboriginals for defence and conquest, were a major trade item. The fur trade intensified the competition and warfare between aboriginal groups, who fought each other to retain a larger share of the market. Consequently, the harvest of furs, previously conducted to meet limited needs for food and clothing, accelerated to unsustainable levels, and populations of some fur-bearing species crashed.

Figure 2.2 The fur trade, a joint venture between Europeans and aboriginal peoples, dominated the North American economy for two centuries. Source: Archives of Manitoba

As aboriginal and European traders moved further into the hinterlands in search of more furs, the trade spread north, west, and south, the Canadian boreal forest providing the richest pickings. The British entered the trade through Hudson Bay, with a royal charter issued to a company bearing the same name. The Dutch, later displaced by the British, pursued the trade up the Hudson River and into the continental interior. The search for furs led European traders across the continent to the Pacific, where maritime trading posts had already been established.

The fur trade, especially in Canada, evolved into a form of partnership between aboriginal and European peoples. The traders had no interest in settling their trading regions and in most cases actively discouraged settlement. They came, took what they wanted, and returned home. The aboriginal people remained in control of their lands and, after the initial over-harvest of fur-bearing species in the east, continued their fur harvests on a more or less sustainable basis. Because habitat remained intact, animal populations rebounded quickly when trapping pressures declined.

The fur industry, the continent's first real forest industry, lasted for more than two hundred years as the dominant economic activity throughout most of North America. It provided the incentive and much of the financial support for European exploration, conquest, and settlement of the continent. Eventually, however, it was displaced from the east by European settlement.

Settlement and Liquidation

Colonization of Canada did not begin until the seventeenth century, some decades after the fur trade commenced. The main thrust of settlement was from France and Britain, which controlled the New England seaboard to the south. There the winters were less severe, and the land less heavily forested, due perhaps to a longer history of aboriginal agriculture. Settlement of Canada was impeded further by resistance from fur traders, who saw it as a threat to their interests.

French immigrants began arriving in larger numbers toward the end of the century, and over the next few decades settlements were established along the shorelines and navigable rivers of

Figure 2.3 Settlement was accompanied by the clearing of native forests, as illustrated in this 1865 photograph by Alexander Henderson of a farm near Grenville, Quebec. Source: National Archives of Canada, PA 181769

Nova Scotia, New Brunswick, and Prince Edward Island, and along the lower St Lawrence River.

The pace of settlement was determined largely by the colonial policies of the French and – after the defeat of Montcalm on the Plains of Abraham in 1759 – the British government, which devised various schemes for granting land to settlers. Invariably these schemes included provisions for the clearing of land to make way for agricultural endeavours.

Following the British conquest, King George III issued in 1763 a royal proclamation which, among other provisions, granted North American aboriginal people certain property rights, albeit of an ill-defined nature. Ever since, this proclamation has been central to the legal basis of aboriginal rights and land claims and, as such, has been inextricably related to questions of forest tenure and forest land-use policies.

As had become established practice in the New England colonies, the French and British in Canada instituted what were known as "broad arrow" policies. Trees suitable for shipbuilding such as oak for planking and white pine for masts and spars were reserved to the Crown and marked with arrows blazed into their trunks. Officials were appointed to select areas of forest for reservation and to enforce the laws. The first of these, known as the "Surveyor General of His Majesty's Woods," was appointed in Nova Scotia in 1728. Severe penalties were imposed on anyone cutting these trees without permission – an unpopular policy that, in New England, helped provoke the American Revolution against Britain. Apart from the regulations devised to protect Newfoundland forests from overuse by fishermen, the broad-arrow laws were the first forest laws in North America and constituted the first attempt to regulate forest use. They were not in any sense conservationist in intent but were typically colonial in character – implemented and enforced for the benefit of the contemporary colonial powers, not for the future advantage of those living in the colonies.

The advent of settlement created serious and mostly negative impacts on the forests. For almost two centuries, settlers were heavily dependent on forests for timber. Their houses, barns, and other buildings were made of wood, initially in the form of hewn

Figure 2.4 The use of wood for building, fuel, and industrial production, as illustrated in this photograph of a logging camp in Halifax County, Nova Scotia, consumed large amounts of timber during the settlement of Canada. Source: Nova Scotia Archives and Records Management

logs and later from lumber milled in the water-powered sawmills that quickly appeared in every settlement. As fields were cleared and livestock acquired, fences were needed. Until barbwire was invented in 1873, these fences were made of wood, usually of split rails. It required eight to ten thousand rails to surround a sixteen-hectare field.

Until the mid-nineteenth century when coal became readily available, settlers and their descendants – not to mention the growing urban populations – were entirely dependent on wood for fuel. Canada's cold winters created a demand for fuel far above that in the New England colonies. An average rural household might consume thirty cords of firewood a year. In 1870 wood provided 90 per cent of Canada's annual 200 trillion BTU (British Thermal Unit) energy needs.

The most daunting task confronting an eighteenth or nineteenth century settler in Canada, apart from those on the open

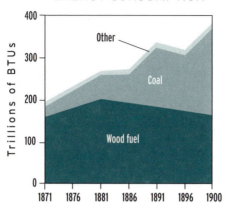

ENERGY CONSUMPTION

Figure 2.5 After 1870 coal supplied a growing portion of Canada's energy needs, although the consumption of fuel-wood declined slowly. Source: *Historical Atlas of Canada*

prairies, was clearing the land of trees to make way for creation of a home site, the growing of crops, and the raising of livestock. Trees had to be felled, bucked, and piled for burning by hand. Draught animals were used to skid building logs or to haul firewood, but sleds and wagons were hand loaded. Clearing a hectare of forest land might take a healthy, vigorous settler two to three months. Only then could he begin to grow crops between the stumps, which might take years to rot away.

In spite of the difficulty of this task, there was a general tendency to clear as much land as possible, leaving forested only those portions unsuitable for agriculture. Occasional voices of dissent were raised, advising the retention of woodlots to provide for future timber and fuel needs and to maintain water tables, provide wind breaks, and maintain wildlife habitat. But generally settlers chose to maximize their individual agricultural potential, obtaining what timber and fuel they needed from forests elsewhere. Landscape portraits of the St Lawrence Valley in the late eighteenth century indicate that forest liquidation was virtually total in settled areas, even at this early stage.

Given the amount of back-breaking labour involved in clearing land, it is understandable that settlers resorted to the use of fire whenever possible. It was much easier to remove trees if the underbrush was burned, and in some cases it was a simple task to eliminate the forest with fire.

In fact the burning of wood provided early settlers with one of their few sources of cash income. Wood ash was used to make potash, a potassium compound used for fertilizer. In 1870 there were more than five hundred asheries operating in Canada, exporting more than forty thousand barrels of potash a year.

The problem with the use of fire was the difficulty in controlling it. When conditions were right for burning, the forest was often dry enough for fires to escape. Wildfires begun by settlers clearing land caused the first great wave of forest destruction. The most spectacular fire of this origin was the 1825 Miramichi fire, which began when several settlers' clearing fires escaped and joined into a gigantic conflagration that burned one-quarter of New Brunswick, killing two hundred people and burning several settlements along the Miramichi River. It was notable only for its size, not its source of origin. Apart from the loss of life and property, at the time the Miramichi fire was not looked upon as much of a calamity. The forest itself was considered to be of little value, and the fire did succeed in clearing a large area of land.

This response was a reflection of the prevalent attitude about forests in colonial Canada. At best, conventional wisdom was

Figure 2.6 Until late in the nineteenth century, the burning of forests was widely considered advantageous. Source: Canadian Forest Service

ambivalent when it came to consideration of forests. At worst, they were seen as evil impediments to civilization that should be eliminated as quickly and efficiently as possible. Most European colonists brought with them to the so-called New World a cultural antipathy to forests. Their myths, legends, and literature abounded with images of forests as dangerous places. These stereotypes were reinforced in North America, where settlers' livestock were under constant threat from forest predators such as bears, wolves, and cougars. Among the first bills to be passed in every colonial or provincial legislature was one providing for bounties to be paid on these and other undesirable creatures of the forest.

As well, the century-long war between early French colonists and the Iroquois, which included aboriginal attacks on isolated settlers, fuelled the early Canadian fear of forests. From the settlers' perspective, forests harboured uncivilized, murderous savages, and so forest liquidation was widely considered a socially desirable objective until the late 1800s. During this period, the primary agent of that liquidation – after the settler – was the lumberman. The two were often the same person. Typically, settlers worked in logging camps during winter months and returned to their farms for the summer growing season.

Utilization of timber on a commercial scale did not begin in colonial Canada until just before the American War of Independence in the late eighteenth century. Before then, most timber removed from the forests was for local use. There was some shipment of masts and spars by both the French and English, and some trade with the West Indies; otherwise, commercial exploitation of the forests was restricted by the broad-arrow laws and the high cost of shipping to Europe. The French obtained most of the timber products they needed from the Baltic region, as did the British when not blockaded from this source by the French. Additional British demand was met from the New England colonies.

The American Revolution had a dramatic impact on Canada's forests. First, it forced British naval authorities to turn to Nova Scotia and New Brunswick for naval timbers, launching a timber trade that exists in modified form to this day. And it created a massive exodus of loyalists from the United States into British North America, among whom were many experienced New England lumbermen with the skills, capital, and market connections to produce and export forest products on an industrial scale.

The War of 1812 between Britain and the U.S., and a concurrent British conflict with France, created new demands for Canadian timber. The French blocked the British supply of timber from the Baltic, increasing British reliance on timber from its Canadian colonies. This demand was sufficient to push the newborn forest industry up the St Lawrence River into the Great Lakes and up the Ottawa River into the vast pine forests of Upper and Lower Canada.

Timber for the European – primarily British – trade was hand hewn in the woods into squared timbers and spars. Great rafts of these products were floated downstream to deep-water locations such as Montreal, Quebec, Miramichi, and Saint John, where they were loaded onto ships for transport across the Atlantic. There the timbers were sawn into the desired dimensions. This was an enormously wasteful use of wood, but often highly profitable all the same. The most desired species were white and red pine, and the pursuit of these species resulted in the expansion of the industry

MAJOR WOOD PRODUCTS

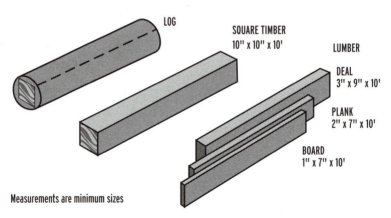

Figure 2.7 Square timber, hewn from logs in the forest, was the main product of the early timber trade. Deals, planks, boards, and lumber sawn from logs or square timber in sawmills gradually replaced the square timber trade. Source: *Historical Atlas of Canada*

along the St Lawrence into the Ottawa Valley and the eastern Great Lakes.

Square-timber exports from British North America increased from five million cubic feet in 1810 to a peak of forty million cubic feet in 1846, when the market collapsed because of over-production. From this point, square timber production slowly declined and was eventually replaced by sawn lumber.

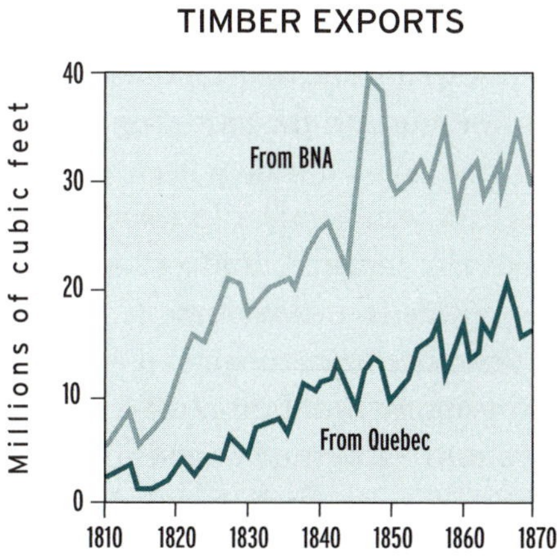

TIMBER EXPORTS

Figure 2.8 The growth of timber exports from eastern Canada, beginning during the War of 1812, increased throughout the nineteenth century. Source: *Historical Atlas of Canada*

The rapid development of the colonial forest industry required new regulations to provide lumbermen with access to timber, beyond what they could obtain under the broad-arrow policies. During the first half of the nineteenth century, a series of laws and regulations were adopted in Nova Scotia, New Brunswick, and Upper and Lower Canada (the last two united as the Province of Canada in 1841) that laid the foundations of forest policies that would prevail throughout Canada until the present time.

In 1816, for instance, Britain granted the New Brunswick colonial administrators authority to regulate the timber industry. One of their first acts was to introduce a timber licence requiring loggers to operate in a specified area and to pay a royalty of one shilling a ton on timber harvested.

Rights to the resources on Crown lands, and to the revenues generated from them, passed to the colonies as they assumed responsibility for the costs of administering civil government, beginning with New Brunswick in 1837. These developments established two principles in Canadian forest policy. First, forest resources were placed under the jurisdiction of the colonies and later were passed to the provinces. Second, revenues from the sale of forest resources were used to finance government operations.

As the centre of timber production moved westward into the Ottawa Valley and the Great Lakes basin, the united colonies of Upper and Lower Canada adopted similar policies. The Crown Timber Act of 1849 confirmed the policy of retaining Crown ownership of forest land, while providing increased security of tenure to lumbermen on their timber licences. The primary objectives of these policies were to clear forested land for agricultural settlements, make raw material available to the growing timber industries, and provide a steady flow of revenues to colonial administrators. Revenues from the sale of timber on Crown land did not require the approval of legislative councils as did taxation or most other sources of government revenue, and were less politically contentious. An additional objective was to clear forested land for agricultural settlement.

The basic mechanism of the new policies was to facilitate timber harvesting through a variety of temporary leases and licences, in contrast to policies adopted in the newly constituted United States, which transferred large areas of forested land into private ownership. A royalty or stumpage fee was levied on the timber. Title to the land remained with the Crown, which was able to grant it to settlers after it was cleared of merchantable timber.

Three pivotal Canadian forest policies were enshrined in law during this period: Crown ownership of forest land, the sale of timber from Crown land, and a dependence on timber sales for public revenues. These policies formed the basis of a unique Canadian partnership between timber-based forest industries and colonial (later provincial) governments that controlled the majority of forest land. It was a set of policies beneficial to both parties. Governments received revenues which they could spend on measures popular with the electorate, and lumber producers were able to avoid the costs of land ownership – interest and taxes – borne by their competitors in Europe and the United States, where most timber was obtained from private land. Additionally, the revenue-collection system required payment for timber only after it was cut and sold, reducing further the capital requirements of the industry. These policies have endured in most of Canada for more than 150 years and only recently have come under serious re-examination.

With a growth in demand for lumber in the U.S. during and after the American Civil War, the timber industries in Atlantic and central Canada expanded to include a sawmill sector. Large-scale industrial use of forests in British Columbia began with the establishment of sawmills at Port Alberni on Vancouver Island and in Burrard Inlet on the lower mainland in the 1860s. A network of canals and railways carried lumber from Ontario and Quebec mills to the American market centred in New York. Lumber from the Atlantic provinces flowed to Britain, the West Indies and the U.S., with B.C. lumber finding markets in California and around the Pacific Rim.

Figure 2.9 Timber raft on the Ottawa River - from the 19th century onward, huge rafts of square-hewed lumber were floated down the Ottawa and St Lawrence Rivers for export to Britain. Source: National Archives of Canada, PA 139334

In the B.C. Interior and the Yukon, other developments heavily impacted regional forests when a series of gold rushes occurred during the last half of the century. Tens of thousands of gold seekers poured into areas such as the southeastern Kootenays, the

Cariboo, and the Klondike. Large amounts of timber were required for mining operations, the building of townsites, and fuel. Little care was taken to avoid fires, and many were set deliberately because it was easier to locate mineral deposits when the forest cover was removed.

Yukon forests were utilized extensively during the Klondike gold rush, beginning in 1897. Vast amounts of lumber were needed to meet the usual type of construction needs, as well as to build a fleet of 250 large steamboats that plied the territory's navigable rivers, providing the region's only form of transportation. These wood-fired boats burned huge quantities of cordwood. A single voyage over the 1,440-kilometre return trip from Whitehorse to Dawson consumed more than 250 cords of wood. Even larger amounts were burned for heating houses and for thawing the ground to permit mining to continue through the winter months.

Figure 2.10 The Klondike riverboat *Gleaner* taking on fuel-wood in the 1890s. Source: Vancouver Public Library 13366

Canadian Confederation in 1867 brought to a close the colonial era of forest liquidation and exploitation. This historic political event coincided with the beginning of a rapid period of industrialization which was to have an even larger impact on the new nation's forests.

CHAPTER 3

INDUSTRIALIZATION OF THE FOREST

The industrialization of Canada, largely with the widespread adoption of steam-powered machinery, took place through the latter half of the nineteenth century and the first quarter of the twentieth. It impacted forests directly and indirectly in various ways.

Mechanization of the forest industry itself greatly speeded up the pace of logging and milling operations, consuming larger volumes of timber and affecting an increased area of forest. This development coincided in central Canada with the logging-out of stands along rivers suitable for the driving of logs. The availability of steam locomotives enabled the extension of logging operations into more remote but well-timbered locations such as the Canadian Shield and the areas beyond British Columbia's coastal fringe.

The first mechanized logging equipment consisted of railways, cable-logging systems (used almost exclusively in B.C.) to yard logs to rail lines, steam haulers used to pull log sleighs along ice roads, and amphibious warping tugs, called "alligators," used to tow logs across lakes in the Canadian Shield. These machines increased the loggers' productivity, enhancing the competitive ability of the industry in export markets. They also increased the impact of logging on the forest. In West Coast forests, for instance, pre-industrial logging was done with oxen and horses. It was a form of selective logging that took only the most commercially valuable timber, leaving the rest. In many cases this produced second-growth forests dominated by diseased or deficient trees. In others, healthy, well-stocked stands of young trees were left.

With industrialization and the adoption of cable-logging sys-

Figure 3.1 Wood-burning locomotives, such as this Shay working at Lang Bay on the B.C. mainland coast, consumed large volumes of wood for fuel to haul logs, and even larger volumes in the fires they started.
Source: Campbell River Museum and Archives photo 8719

tems such as high lead and skyline methods, clearcut logging became the common practice. The evolution of cable systems, combined with logging railways, led to the progressive clearcutting of entire watersheds from the estuaries to the headwaters. Trees not utilized were knocked down and left. This form of logging left large volumes of slash which often burned, creating intense fires that damaged soils and spread into adjacent, unlogged stands.

Steam-driven sawmills replaced water-powered mills, enabling construction of larger mills in urban centres and port cities. Construction of a continental rail system accelerated sales of Canadian lumber in expanding U.S. markets. The move to steel-hulled steamships from wooden sailing ships, on the other hand, reduced the demand for ship timbers after the peak construction period for wooden ships (1875–80).

Probably the most significant form of industrialization in Canada during this period was the building of railways, particularly the transcontinental lines. The influence of railways on forests was severe and varied. Construction of rail lines and their related facilities consumed immense quantities of timber. The 4,800-kilometre-long Canadian Pacific Railway (CPR) from Montreal to Vancouver, begun in 1880, was the largest of such undertakings. It required two thousand wooden ties for every kilometre of track – all of which had to be replaced every three or four years.

Hundreds of bridges were needed, all built of wood until funds were available to replace them with steel structures. The larger bridges used two million board feet or more of milled timbers. After the first difficult year of operations, dozens of snow sheds were built through the mountainous regions of B.C., consuming tens of millions more board feet of lumber. Telegraph lines, wooden boxcars, stations, roundhouses, and other buildings consumed even greater quantities. To provide it, scores of sawmills were built at strategic locations along the route, tapping timber resources across the continent. The prairie sections were built largely with timber logged in northwestern Ontario, the mountainous sections of eastern B.C. with wood from mills on the Columbia River, and the western sections from coastal mills.

The first transcontinental line was only the beginning. Other main and branch lines followed, all consuming comparable volumes of timber. Until coal became widely and readily available, most of these rail operations were fuelled with wood. By the end of the nineteenth century, almost 32,000 kilometres of railway lines had been completed, and by the end of World War I, more than 80,000 kilometres.

Although land grants for construction of railways were less common in Canada than in the United States, they were employed and provided alternative access to timber. Construction of the CPR entailed a grant of ten million hectares to the company, most of it unforested. However, in southeastern B.C., more than three million hectares, most of it forested, was granted for railway construction, and another three-quarters of a million hectares of the most valuable timber land in the province granted for construction of the Esquimalt & Nanaimo Railway on Vancouver Island. The trees on these lands provided a primary supply source for industry for several decades. Another block running for thirty-two kilometres on either side of the CPR route

Figure 3.2 Building the Canadian Pacific Railway required the use of large volumes of timber in bridges and snow sheds, such as these in the Selkirk Mountains of B.C. Fires, like those that burned the timber stand above the snow sheds, were common along this and other railways. Source: Whyte Museum of the Canadian Rockies photo, v653/N64-255

through B.C., as well as a large block in the Peace River district, was given to the federal government by the province to finance railway construction. In this manner the country's forest resources were used freely to construct and finance its industrial infrastructure.

The third consequence of industrialization for forests was the fires started by steam machines burning wood and coal. These fuels, when used in locomotives, donkey engines, and other mechanical devices brought into the forests, belched sparks from their exhaust stacks, igniting fires in the logging slash. Steel railway wheels running on steel rails threw sparks, and steel logging cables yarded over slash generated more fires.

The railways were perhaps the worst offenders. By World War I, a substantial percentage of the forests bordering the tracks had been burned. J.H. White, the first graduate of the University of Toronto's faculty of forestry, reported in 1912: "From Sudbury to Port Arthur, generally speaking, the country along the [Algoma Central] railway has been burned at one time or another for the entire distance of 550 miles. Not much has escaped except the spruce swamps. The burned areas have been partially recovered by temporary stands of poplar, white birch and jack pine, either pure or in mixture. But to a vast extent the country has been burned so repeatedly that there is nothing left but bare rock."

Later in his report White mentions that "fire in the last 50 years has reduced the pine area north of Lake Huron by one-half." Similar conditions prevailed throughout the rest of the country. This situation was exacerbated by several factors. Many railways were chartered by the federal government while forest regulations were administered by the provinces, making it difficult to develop and enforce protection rules. The basic cause, however, was the prevalent notion of forest liquidation. Well into the twentieth century the sentiment was still widespread that forests were an impediment to development and settlement, and their eradication was acceptable, if not desirable.

Industrialization, and particularly the development of railways, dramatically accelerated the pace of settlement, especially in the southern boreal region of the prairie provinces. The settlement of this area was responsible for the most extensive liquidation of forests in Canadian history. While the first decades of the nineteenth century had seen the virtual extirpation of the deciduous forest region in southwestern Ontario, the early years of the next century witnessed a more extensive forest liquidation along the southern fringe of the boreal, from the Ontario-Manitoba border west to the Rocky Mountains. This settlement was greatly facilitated by railway development.

Settlement of the prairie parkland forests also involved the largest assault ever on a forest wildlife population. Wildlife habitat in the St Lawrence Valley and the Great Lakes region was seriously disturbed, threatening and endangering the existence of many species occupying the region's forests. The impact of industrialization and subsequent settlement of the prairies on the region's prolific buffalo herds was far graver. Ironically, the first significant freight item of the new prairie railway network was the bones of the slaughtered buffalo, transported east to be ground into agricultural fertilizer.

Figure 3.3 Completion of the transcontinental railway led to the destruction of the vast herds of buffalo that populated the prairie grasslands. Their bones were later collected for transport by rail and made into fertilizer. Source: Canadian Pacific Railway Archives NS.6679

Industrialization also hastened the growth of the timber industries. The rise in population that accompanied it, as well as the industrial processes themselves, increased the domestic demand for timber. There was also growth in international timber demand and in exports. The United States, which had cut out its northeastern forests, required vast volumes of timber to facilitate the rapid opening of the West to settlement after the Civil War.

The value of Canadian lumber shipments to the U.S. rose from $9.5 million in 1873 to almost $150 million in 1920.

This period saw a pronounced shift in the centre of Canadian lumber production, especially lumber for export. By the end of World War I, the eastern lumber industry had liquidated the bulk of its available sawlog supply, particularly in the Atlantic provinces, and began to convert its forest economy to one based on pulp and paper. Lumber production shifted to B.C., beginning with completion of the CPR in 1885, at which time the province's large coastal mills cut 75 million board feet a year. By 1906 coastal mill production had reached 525 million board feet a year, a rate of expansion that would be sustained, with brief interruptions, until the Great Depression of the 1930s. This increased activity was aided by the opening of the Panama Canal in 1914, which provided B.C. producers with economic access to European markets.

By the early twentieth century the forests of southern Canada were under immense and growing stress from several directions. In some areas, especially the prairies, they were being liquidated to make way for agriculture-based settlement. In others, primarily the Atlantic provinces and the accessible forests of Ontario and Quebec, a rapidly expanding timber industry was depleting the commercially valuable forests. And across the country, fires, some resulting from natural causes but most through human carelessness, were burning huge areas of forest. The state of the Canadian forest in the early years of the century was probably at its lowest point since the glaciers had receded.

State of the Forests 1900

It was – and still is – difficult to know with any degree of accuracy the condition of the country's forests around the time of the First World War. Inventory data were sparse or non-existent, so most descriptions were at best educated guesses.

The Commission of Conservation, established by the federal and provincial governments in 1909 and assigned the task of providing information on natural resources upon which policy could be based, conducted two provincial forest inventories. The first, a crude inventory of Nova Scotia's forest conditions, concluded, "Seventy per cent (of the province) is actual or potential forest land, very much of which is now in poor condition, but is capable of restoration. At the present rate of cutting, the merchantable timber will be exhausted in from twenty to twenty-five years." Much of Nova Scotia's productive forests had been stripped of pine, after which they had been burned, often repeatedly. Large areas of the province's original forests grew in a thin layer of soil that overlay impermeable rock. Large areas of natural rock barrens had increased substantially after recurrent fires burned away the soil.

A similar situation existed in Newfoundland, with the creation of extensive man-made barrens, especially around fishing settlements. Part of the problem in Newfoundland originated with the early establishment of commonly owned land in a strip 4.8-kilometres wide strip along the coastal margin. This was intended to provide the fishing industry, and fishing community, with a perpetual supply of wood for construction. But belonging to no one, it was misused by all.

In Quebec most of the forested land along the St Lawrence and lower Ottawa River had been cleared for settlement, which by then was extending to areas unsuitable for agriculture. The industrial assault on the province's timber resources had also taken a toll, as an 1887 report to the federal minister of agriculture indicated: "In a very short time since the beginning of the century we have over-run the forests picking out the pine, and we have impoverished them to a serious extent. There still remains to us a great deal of spruce and second-rate pine, which for generations to come will be in excess of our wants, if we are careful, but the really fine pine is getting very scarce and inaccessible, and I feel that we must prepare for a serious falling off."

The situation in Ontario, as described in a 1899 Royal Commission report on forest protection, was similar:

Between the need of the farmer to clear his land for farming, and the wants of the sawmill man, much land in [Southern Ontario] has been cleared of trees that would have been far better kept in permanent forest. It was a very shortsighted policy that removed the trees from the hillsides, and the hills, from broken and uneven land, and along the head waters of the streams, allowing the blasting and drying wind an uninterrupted sweep across the country, allowing the washing of the soil off the hillsides into the valley below to the detriment of both, and removing the causes that kept the streams and springs perennial.

Long observation and experience has demonstrated that, aside altogether from the needs for timber and fuel, the welfare of the community requires that 20 to 25 percent of the total area of a country should be tree covered. Instead of 25 percent of forest, some of the counties in [Southern Ontario] have not over 5 percent, and this in such scattered clumps of scraggy trees as to be of little use for climatic or water supply purposes. Nearly every spring the Grand River overflows its banks and causes heavy damages at Brantford and elsewhere. This stream flows through Brant, Waterloo and Peel. None of these counties have over 15 per cent of wood land – Brant has only 7 percent, Peel about the same, Waterloo has about 13 per cent.

The most exhaustive forest inventory conducted in Canada in this period was undertaken by the Commission of Conservation in British Columbia, which published a report in 1918. It began with an overview of the province's forests:

It has been found that, of the total land area of the province, 355,855 square miles, approximately 200,000 square miles is incapable of producing forests of commercial value . . . Deducting the potential agricultural land, say 20,000 square miles, from the land capable of producing commercial timber, there is 135,855 square miles of absolute forest land which should be devoted permanently to forest production.

The timber on about 100,000 square miles, or two thirds of the land once forested, has been totally destroyed by fire, and on over half of the remaining 55,855 square miles the timber has been seriously damaged. Using the timber still standing as a basis, it is estimated that the province has lost, through forest fires, at least 665 billion board feet measure. When one considers that the total stand of saws material in the whole Dominion probably does not greatly exceed this amount now, the seriousness of this loss, which can be attributed very largely to public carelessness, becomes apparent.

The situation was not much different in the Yukon Territory, where the most valuable timber, found along the navigable rivers, had been cut during the gold rushes, and much of what remained had been burned in wildfires. The remainder of the central and northern boreal forest, however, remained untouched. It, along with most of the interior B.C. forests, contained vast areas of unmerchantable timber – either because of its small size or because of its species. Large areas of British Columbia, for example, were covered with almost pure lodgepole pine stands, which would not be considered merchantable and utilized for another fifty years.

These and other reports, combined with a growing public awareness, began to challenge a long-standing myth that Canada's forests were inexhaustible. The threat of future timber famines, the growing incidence of floods, erosion, and dust storms, and the depletion of some once-abundant wildlife species eventually became a matter of public concern. Gradually, the voices of the country's small but growing contingent of forest conservationists began to be heard.

CHAPTER 4

THE RISE OF FOREST CONSERVATION

The North American conservation movement was based on ideas and concepts that began to appear long before the end of the nineteenth century, when forest liquidation in Canada had become widespread.

Writing from his office in the House of Commons overlooking the Ottawa River, Prime Minister John A. Macdonald noted with alarm in 1871: "The site of the immense masses of timber passing my windows every morning constantly suggests to my mind the absolute necessity there is for looking into the future of this great trade. We are recklessly destroying the timber of Canada and there is scarcely a possibility of replacing it."

The theory of forest conservation came to Canada from older societies in Europe and Asia that had experienced the consequences of deforestation earlier in their histories. There were two primary sources of the new thinking about forests for Canadians: French-speaking intellectuals, politicians, and lumbermen in Quebec were heavily influenced by conservationist thought in nineteenth century France. English-speaking Canadians received their ideas primarily from Germany, via the United States.

The first North American forestry schools opened in the U.S. in the 1890s and taught German-style forestry. Several prominent U.S. foresters, including Gifford Pinchot, the first American-born forester, were trained in Germany. The first English-speaking Canadian foresters were trained at these American schools, which were hotbeds of conservationist thinking. Bernhard Fernow, a German forester, started the Cornell University school of forestry, served as the first U.S. chief forester, and was the founding dean of Canada's first forestry school, which opened in 1907 at the University of Toronto.

The fundamental concept upon which conservationist doctrine was built was that of the permanent, or perpetual, forest. It was in direct opposition to the prevalent North American concept of forest liquidation that lay behind the widespread deforestation that was alarming large numbers of people by the dawn of the twentieth century. Science-based management of the forest, it was postulated, would provide society with a wide range of benefits, including fuel, timber, clean water, fish, wildlife, and recreation. Conservation, one observer noted, entailed the use of "foresight and restraint in the exploitation of the physical sources of wealth as necessary for the perpetuity of civilization, and the welfare of present and future generations."

A key component of conservationist thinking was that planning and management of forests and other natural resources should be undertaken by scientifically trained professionals, not by politically appointed officials, as was common practice at the time. Experts employing technical and scientific methods, rather than politicians or industrialists and other forest users, should be in charge of forests.

In Canada the conservation crusade was led by a loose coalition of political leaders, lumbermen and a handful of the country's first forest academics and professionals. James Little, a Montreal lumberman with large timber holdings and mills in southwestern Ontario and Quebec, was one of the earliest and most articulate proponents of conservationist thinking in both Canada and the U.S. He argued persuasively for protection and retention of forests at a time when such ideas were considered radical and unnecessary.

Sir Henri Joly de Lotbinière was the head of a family that had owned 35,000 hectares of forest land on the St Lawrence River west of Quebec since 1673. Perhaps the country's most influential conservationist, he introduced European-style scientific forestry to his lands at an early date and demonstrated in practical ways the economic wisdom of managing forests in perpetuity. Trained as a lawyer and elected to the Assembly of Canada in 1861, he was an ardent proponent of Confederation. In 1900 he served as the founding president of what quickly became the country's pre-eminent conservationist organization, the Canadian

Figure 4.1 Sir Henri-Gustave Joly de Lotbinière (1829–1908) was a leading North American conservationist and a member of the Quebec Assembly, the Assembly of the Province of Canada, and the Canadian House of Commons. As lieutenant-governor of British Columbia, he introduced the concept of forest conservation to western Canada. Source: Forest History Society

Forestry Association. Later that year he was appointed lieutenant-governor of British Columbia, a position that enabled him to introduce conservationist ideas into the formation of that province's early forest policies.

Sir Wilfrid Laurier, who became prime minister of Canada in 1896, was another dedicated conservationist and hosted the country's first forest congress in 1906. This congress defined conservationist forest policies for decades to come and marked a turning point in public opinion and official policies regarding forests. It was attended by most of the continent's forest conservationists.

In his address to the congress, Gifford Pinchot carefully defined mainstream conservationist theory, distinct from a preservationist form of the movement which had gained influence under John Muir, a prominent American naturalist who founded the Sierra Club in 1892. Muir campaigned successfully for the creation of parks as permanent reserves where industrial activity was prohibited. He saw wilderness as a place of rejuvenation for people worn down by the modern industrial world: "In God's wildness lies the hope of the world – the great fresh unblighted, unredeemed wilderness. The galling harness of civilization drops off, and wounds heal ere we are aware."

The first Canadian parks, beginning with the creation of Banff National Park in 1887, were established during this period, for similar motives. H.R. MacMillan, a young Ontario forestry student studying at Yale, wrote about the role of parks after spending a summer working on the Turtle Mountain forest reserve in Manitoba:

The aesthetic value [of the forest] cannot be over estimated. By the determined action of the earliest settlers who drove past three hundred miles of arable prairie to reach the forested hills, by the ever increasing Summer camps, an appreciation of the presence of the forest is amply

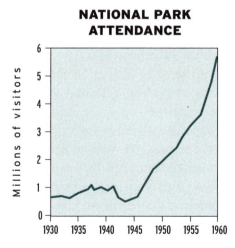

Figure 4.2 National park attendance soared with the return to prosperity and increased availability of automobiles after World War II. Source: *Historical Atlas of Canada*

demonstrated. Even now there are from fifty to one hundred people who spend the Summer camping in Turtle Mountains. As the population increases, days of ease grow nearer, and the active, booming age of realism gives way to a more human sentiment of appreciation this number will yearly grow larger. Here they send their children during the long Summer vacations, here they entertain their visitor, instead of remaining cooped up in the little sun-baked red-hot prairie towns, and here, on the timber reserve, is the one spot, which all citizens of such favoured localities value most highly.

Pinchot – as did others at the 1906 congress – took issue with Muir's preservationist position: "We must put every bit of land to its best use, no matter what that may be – put it to the use that will make it contribute most to the general welfare . . . Forestry with us is a business proposition. We do not love the trees any the less because we do not talk about our love for them . . . use is the end of forest preservation and the highest use." Bernhard Fernow reinforced these sentiments, declaring, "Forests grow to be used. Beware of the sentimentalists who would try to make you believe differently."

Prime Minister Laurier ended the congress with a ringing challenge to delegates: "I desire every man in this audience as he goes away to his home and to his own avocation to become a missionary in the work of forestry." The Laurier government's Forestry Branch led the adoption of conservationist principles, hiring most of the country's trained foresters to administer the lands under its jurisdiction.

A third perspective, distinct from those of both Pinchot and Muir but, like them, still within the conservationist ambit, was expressed by Aldo Leopold, a contemporary of MacMillan at the Yale forestry school. For Leopold, writing several decades later, conservation was "a state of harmony between men and land. By land is meant all of the things on, over, or in the earth. Harmony with land is like harmony with a friend; you cannot cherish his right hand and chop off his left. That is to say, you cannot love game and hate predators; you cannot conserve the waters and waste the ranges; you cannot build the forest and mine the farm. The land is one organism. Its parts, like our own parts, compete

with each other and co-operate with each other. The competi-
tions are as much a part of the inner workings as the co-opera-
tions. You can regulate them – cautiously – but not abolish
them."

Over the decade leading up to World War I, the conservationist creed found concrete form in many parts of the country. The
federal government, which had already created a series of forest
reserves in the territories under its control to prevent settlement
and forest liquidation, passed legislation to permit the active scientific management of the reserves under professional direction.
The country's first superintendent of forestry, Elihu Stewart, initiated a tree-planting program on the prairies to provide settlers
with fuel-wood, building materials, and protection from winter
winds.

Implementation of conservationist principles required a refor-
mulation of the industry-government partnership by the inclu-
sion of scientifically trained professionals. Although they would
be employed by the original members of the partnership, the primary function of professional foresters in this relationship was to
ensure that forests were utilized in a manner that would provide
for their future well-being.

Following establishment of the University of Toronto forestry
school in 1907, two more schools opened, one at Fredericton in
1908 and the other at Laval University in 1912. Initially they were
staffed by professionally trained foresters from the United States
and Europe and by some of the first Canadians trained as
foresters outside Canada. The graduates of these schools were
quickly employed in the forest administrations of the federal and
provincial governments.

British Columbia's forest economy was less developed than
that of the central and eastern provinces, and therefore it lacked
their long-vested interests and entrenched legislation. The
province passed a new Forest Act in 1912, based on recommendations of a royal commission on forestry. Both the commission
report and the new legislation were heavily influenced by conservationist thinking, much of it brought to the province by
Lieutenant-Governor Joly de Lotbinière. Professional foresters
from the newly minted U.S. Forest Service, itself an embodiment

of conservationist principles, played a central role in planning the province's new forest act. On the advice of the continent's leading conservationists, MacMillan, then employed in the federal forest service, was appointed as B.C.'s first chief forester.

In many ways MacMillan epitomized the Pinchot strain of conservationist thinking. In his first report to the B.C. legislature he wrote: "The annual growth of the forests of British Columbia is even now, before they are either adequately protected from fire or from waste, certainly not less than five times the present annual lumber cut . . . What is not cut is wasted in the end. It is not merely advisable to encourage the growth of our lumber industry until it equals the production of our forests – it is our clear duty to do so, in order that timber which otherwise will soon rot on the ground may furnish the basis for industry, for reasonable profits to operator and Government, for home-building and, in the last analysis, for the growth of British Columbia."

What MacMillan understood was that forest use would impress upon people the value of forests and that, unless these forests were valued, they would be neglected and ultimately degraded or, as they had been in parts of Newfoundland and Nova Scotia, reduced to unforested barrens. For foresters like MacMillan, who would one day become the country's leading forest industrialist, there was no inconsistency in the two sets of views expressed. From this point of view, forest use and forest conservation went hand in hand; without a healthy forest industry the forests of Canada would not survive.

The primary thrust of the initial flurry of conservationist-inspired activity was forest protection – protection from settlers through the establishment of forest reserves, and protection from fires through the organization of provincial forest services that took as their primary task fire detection and fighting, under the direction of professional foresters. Campaigns were mounted to convince the public, forest workers, and the aboriginal population that the widespread burning of forests, tolerated or even encouraged in the past, was no longer acceptable.

Another significant achievement of this era, from a scientific perspective, was the beginning of forest research. Forest products laboratories were established in Montreal and Vancouver in 1917, and at about the same time the first scientific research facilities were built. A centre of study of forest insects was opened at Vernon, B.C., and in collaboration with the military, a major forest research laboratory was built at Petawawa, Ontario. During the same period, E.J. Zavitz, a forestry professor at the Ontario Agricultural College in Guelph and later the province's chief forester, began a reforestation program in the badly degraded forest lands north of Lake Erie, the first such rehabilitation effort in Canada.

Although the outbreak of World War I had suspended the initiatives fostered by conservationist thinking, the ideas inherent in this powerful social movement laid the groundwork for a massive expansion of industrial forest use, encouraged by the economic revival that occurred after the war. In no small part this was because investors were able to proceed with confidence that adequate timber supplies would exist to feed new mills.

The excitement and enthusiasm of the conservation crusade and its translation into the world of practical affairs masked an inherent weakness in the doctrine as it was widely conceived at the beginning of the century, a weakness that would help generate public conflict over forest use during the latter part of the century. Pragmatic mainstream conservationists such as Fernow, Pinchot, and MacMillan, along with the protectionist Muir, tended to conceive of human beings as separate from nature, as standing apart from the forests they fought so fiercely to protect. A few early conservationists, especially Leopold, adopted a somewhat different concept that perceived humankind as part of nature. Subtle as these distinctions were at a time when the primary objective was to prevent the liquidation of the continent's forests, they did mark a division in the conservation movement. Meanwhile, the utilitarian thoughts expressed at the 1906 forest congress prevailed.

Foresters would spend many decades grappling with the implications of maintaining permanent Canadian forests while at the same time making full use of them. The focus of their efforts dealt with the question of setting limits to this use. The concept employed to describe the process of establishing limits was known as *sustained yield*, a principle that occupied forest-policy makers and practitioners alike for over three-quarters of a century.

The Concept of Sustained Yield

The momentum of the conservation crusade faltered on the battlefields of Europe. A generation of Canadian foresters, drawn out of the newly formed provincial forest services and the country's three forestry schools, marched off to war, and many of them failed to return. Those who did came home to economic uncertainty and an industry in a state of involuntary transformation.

As the postwar recovery gained speed, timber-based forest industries resumed their expansionary mode. The eastern and central sawmill industry enjoyed a dying gasp of prosperity, using up the last of its sawlogs. It was quickly replaced by a dynamic pulp and paper sector that utilized smaller trees and species unsuited for lumber to feed booming markets in the U.S. Pulp and paper companies acquired most of the cut-over timberlands and harvesting licences as lumbermen slowly moved out of the woods in the Atlantic and central regions.

The impact of the pulp and paper industry on the forest was mixed. It utilized much smaller trees than the lumber industry and consequently harvested much more intensively and extensively. But pulp and paper mills required relatively large investments and this had a tendency to stabilize forest use and encourage longer-term thinking about forests. In some cases the new pulp companies took up the torch of conservation that thus far, within industry, had been carried by nineteenth century lumbermen. Some Ontario and Quebec pulp companies hired professional foresters to oversee management of their leased and private lands and launched reforestation campaigns to provide future timber supplies.

In British Columbia, the sawmill sector enjoyed a postwar boom that was sustained through the 1920s. It took over many of the markets of the declining eastern lumber industry and, with the Panama Canal in full operation, shipped to markets around the world. With completion of a second transcontinental rail line through central B.C. during the war, the lumber sector in the interior grew along with that on the coast, where a pulp and paper sector was also developing.

During the 1920s the second phase of forest mechanization

Figure 4.3 The 1915 drive on Bull River in southeastern B.C. where over-harvesting led to mill closures and abandonment of communities, demonstrating the need for sustained yield policies. Source: author's collection

began. Steam-powered machines were replaced with smaller, less expensive equipment powered with internal combustion engines. Trucks and crawler tractors, the latter used to build roads for the trucks and then skid logs to them, replaced locomotives and horses. The new machines could be purchased by smaller firms and individuals, enabling a more diverse logging sector to reach smaller stands of timber previously inaccessible to railway loggers. The use of gas engines on small portable sawmills, which could be set up at temporary sites in the forest, facilitated harvesting of the remaining isolated stands of sawlogs throughout eastern Canada. They were also used extensively in forests throughout the northern portions of the prairie provinces and the interior of B.C. These mills were very inefficient and in some respects were throwbacks to the early days of hewing square timbers in the woods, wasting vast amounts of high-quality wood in the process.

Figure 4.4 Development of crawler tractors and other mechanized logging equipment enabled logging of previously inaccessible areas such as this near Harrison Lake, B.C., in the mid-1930s. Source: author's collection

In a limited, theoretical sense Canadian forests were now looked upon as a renewable resource. And while the old concept of forest liquidation had given way to the idea of the perpetual forest promulgated by conservationists, in reality little had changed in the way forests were used. During the so-called Roaring Twenties, when many people came to believe economic growth would continue forever, the conservationist idea of maintaining social and economic stability through the restrained, efficient use of natural resources carried little weight. The concepts of restraint and stability have little appeal in times of economic growth; they are most popular during periods of economic decline.

The lessons of forest conservation were not lost entirely, however. The profession of forestry was well established, and although foresters were more often hired by industry for their abilities to provide efficient delivery of logs to mills than to manage new forests, they kept the conservationist concepts alive. The realization that there were limits to the use of the country's permanent forests began to dawn. In 1924 a federal royal commission on pulpwood recommended "that on federal forest reserves, such areas which now present opportunities for intensive management for sustained yield, should be immediately brought under measures to that end."

Although the established provincial forest services were unable to develop the full-scale management functions envisioned by turn of the century conservationists, their existence was assured because of their revenue-raising functions. In this mode the well-established forestry profession was able to develop and exercise forest protection capabilities that slowly reduced the destruction of timber by fire, pests, and disease. Forest reserves, which excluded settlers, were retained, preventing the conversion of forest land to other uses.

The practice of forest management, however, and the concept of limited forest use had largely disappeared from public discourse. These ideas were kept alive in large part by professional foresters and industrialists such as MacMillan, who had formed his own forest company at the end of the war. His conservationist principles were still intact in 1927, when he wrote: "The public, as far as I have seen, has very little understanding of how the government, to whom the duty of forestry is assigned, should practice forestry on public lands. The general assumption is that the practice of forestry by the government on public lands is never going to interfere with the profit or comfort of the person who is thinking about it."

A few years later, with the country wallowing in the depths of the Great Depression, lumber production fell 60 per cent below its 1929 peak, and pulp and paper prices plummeted, bankrupting many of the newly created Canadian companies. Those that survived closed down their reforestation programs. In these circumstances provincial governments were reluctant to enforce the few forest-management regulations then in existence.

Again, MacMillan spoke out against these practices: "It is generally known among the well-informed that the forest is being over cut at a devastating rate in every forest province in Canada, that Canada, an essentially forest country, lags far behind India, the United States, Norway, Sweden, Finland and France in forest policy; and that the forest schools and forest departments in Canada are half-starved and failing to lead or influence a Canadian people, who are still bent on exploitation rather than conservation of their great natural resources." His words fell on fertile ground, and his sentiments were widely shared both within the industry and the forestry profession. As provincial forest revenues declined, others paid attention as well.

In 1930, for purely political reasons, control of forests in the prairie provinces and the B.C. railway belt was transferred from federal to provincial jurisdiction. The governments of some provinces allocated funds to examine more closely the state of their forests. The most exhaustive of these studies was a complete inventory of the forests in British Columbia by the provincial forest service, which indicated that harvesting levels in the province's coastal forests, under management practices of the day, were not sustainable. In an introduction to his 1937 report on the inventory, forester Fred Mulholland described the need to move beyond the current policy of unrestricted timber harvesting: "The exploitation of visible supplies without regard for the future, involving 'devastation' is gradually changed to a policy of forest-management for permanent production of 'sustained yield.' By this policy the forests are managed according to plans designed to secure reforestation, regulate the cut, provide sustained annual yields, and stabilize forest industries."

Mulholland went on to discuss future needs: "It has been evident for some time that the forests with protection only will not continue indefinitely to support the great industries already established, to say nothing of the increased production which is now being actively planned. It is time to institute active measures providing for more successful reforestation. It is also time to prepare for some regulation of the cut so that the great store of virgin timber on the accessible areas may not all be used up before sufficient second-growth is ready to take its place."

He concluded with a warning that was applicable to every province in Canada: "Management for a sustained yield is essential for the permanent prosperity of British Columbia's greatest industry and it demands immediate attention. If it is not introduced before the present large forest revenues have disappeared, it is doubtful if capital will be available for the extensive rebuilding of denuded forests which will then be necessary."

The outbreak of World War II intervened, and the entire Canadian forest sector mobilized to support the Allied effort. Among other lessons taught by the war was that the country's forests were of great strategic as well as economic and environmental importance. Until the United States entered the war, more than two years after it began, and with all of Europe except Britain overrun by Axis forces, Canada was Britain's chief ally. The most critical material contribution from Canada during this period was timber harvested from its forests.

But even before the war ended, steps were taken to heed the warnings of MacMillan, Mulholland, and others. In 1943 the government of British Columbia called a royal commission that was instructed, among other tasks, to address "the establishment of forest yield on a continuous production basis in perpetuity." Over the next few years, six more provinces struck royal commissions with similar terms of reference.

The B.C. commission's report, written by Chief Justice Gordon Sloan, examined the concept of sustained yield in great detail and defined the problem facing the entire country: "At present our forest resources might be visualized as a slowly descending spiral. That picture must be changed to an ascending spiral. Differently phrased, we must change over from the present system of unmanaged and unregulated liquidation of our forested areas to a planned and regulated policy of forest management, leading eventually to a programme ensuring a sustained yield from our productive land area."

Sloan defined sustained yield as "a perpetual yield of wood of commercially usable quality from regional areas in yearly or periodic quantities of equal or increasing volume." The objective of provincial forest policies, he wrote, must be: "To so manage our forests that all our forest land is sustaining a perpetual yield of timber to the fullest extent of its productive capacity. When that is accomplished all benefits, direct or indirect, of a sustained-yield management policy will be realized; providing, of course, that the multiple purpose of our forests is recognized as an aim as important as balancing cut and increment."

Other provincial royal commission reports expressed similar sentiments although their specific recommendations were often muted in deference to various established interests, public and private. A 1947 Ontario commission conducted by Major General Howard Kennedy concluded: "In future Government action, the principle of sustained yield must ever apply. Any other course will spell eventual disaster to many of our existing industries and the communities they support."

In a similar vein a dominion-provincial conference in 1945 concluded "that forests, far from being inexhaustible, are being seriously depleted. There is growing realization of the fact that if our forest industries are to exist and expand, the forest must be handled as a crop and not as a mine, in other words must be managed on a sustained yield basis."

Changes in British Columbia in the wake of the Sloan Commission's report were extensive. The province's forests were divided into a series of sustained-yield units, each with a regulated volume of timber harvest. The broad principle guiding the calculation of these allowable annual cuts was to create forest management units that would eventually produce a sustainable, even flow of timber for the forest industries.

To facilitate this type of management, a new form of lease, called a Forest Management Licence, was created. It constituted a form of partnership between the provincial government, which retained ownership of the forests, and private forest companies, which obtained the right and responsibility to harvest mature timber and manage the forests according to government-approved plans. One of the requirements was that a professional forester oversee the management of these forests. Over the next few decades this type of lease, based on the principle of an industry-state partnership, was adopted by virtually every province in the country.

The sustained-yield policies adopted during the postwar period had a mix of strengths and weaknesses. They both facilitated and were shaped by the tremendous postwar boom that endured for more than thirty years. The economic resurgence was of great consequence to the country's forests. On one hand, it created an enormous demand for industrial timber to feed booming markets within Canada and abroad. On the other hand, the technical developments upon which it was largely based drastically reduced the dependence on wood for fuel.

The rapid mechanization of agriculture after the war reduced the dependency on draft animals which themselves occupied more than one-quarter of all farmland. In addition, the introduction of fertilizers, herbicides, and other innovations increased farm productivity. Consequently, the pressure for continued conversion of forests to farmland eased even as demand for food grew. In time, as productivity increased even more, marginal farmland was abandoned and began reverting to forest.

As it continued, postwar prosperity provided Canadians with increased leisure time and the income to enjoy it. A proliferation of automobiles and roads enabled people to enjoy recreational activities in the country's forests.

To a large extent the increase in demand for timber products in the 1950s and beyond was satisfied through improvements in timber-processing and harvesting technology. Wartime technical improvements such as rubber tires, small engines, and hydraulics underlay a wholesale retooling of the industry. Although the B.C. sector had mechanized between the wars, mostly with steam-powered equipment, the central and eastern sectors still depended heavily on the muscle power of horses and humans to harvest timber. This situation changed rapidly after the war with the introduction of power saws, rubber-tired skidders, hydraulic loaders, and other machines. In the B.C. interior a large lumber sector developed, selling primarily into the U.S. market and utilizing lodgepole pine, a small-diameter species previously considered unusable by industry. The growth of this new sector was greatly facilitated by development of a new type of saw, the Chip-N-Saw, which cut lumber at high speed and great accuracy and, as a by-product, produced pulp chips that provided raw material for a pulp and paper industry.

This is not to say that technical innovations reduced the vol-

Figure 4.5 The area of forest logged annually in Canada increased steadily throughout the twentieth century. Source: "Canada's Forests at a Crossroads"

ume of timber or area of forest harvested, because both those factors increased steadily throughout the twentieth century. The area of forest logged annually in Canada increased from 400,000 hectares in 1922 to the greatest amount ever in 1988 when 1,100,000 hectares were harvested. Under a sustained yield regime, however, the attitude towards the fate of the harvested lands was much different from what it had been. The amount of the annual harvest, the allowable annual cut, was directly affected by the amount of growth in the young forest. The more quickly that harvested land within a sustained yield management unit was reforested, the higher the allowable annual cut. Prompt reforestation of deforested land became a priority.

More important than any other factor, perhaps, prosperity created a public willingness to abandon the liquidationist forest policies of a previous era and adopt new ones restricting current forest use to provide for future needs. The phase of public indifference to forests was coming to an end.

From the industry's perspective, sustained yield policies were a great boon. In most parts of the country, their adoption did not require harvest curtailments and mill closures. On the contrary, they provided a degree of certainty and security about future timber supplies – more often than not at higher harvesting levels – which encouraged increased investment in the industry.

There were, however, some serious flaws in the application of sustained yield concepts throughout Canada. The country's tim-ber-based forest industries have always exported more of their production than is consumed domestically, especially in the case of newsprint. Sustained yield policies were intended and designed to produce an even flow of industrial timber; but demand in the export markets fluctuated, often dramatically. Canada's sustained yield regulatory structure tended to force a fixed volume of timber onto falling markets and limit the amount available in rising markets, increasing the severity of price fluctuations. This produced economic instability in Canada and – at least in the case of the U.S. – irritation abroad.

By limiting timber harvests of some species in some regions, sustained yield policies, especially when combined with an increased capability to prevent and fight forest fires, gradually produced large areas of abnormally old or over-mature forests susceptible to insects and disease. As well, many of these old forests stagnated, and the timber volumes they contained declined. But the biggest problem with forest policies in place across Canada during the postwar era was not because of any-thing inherent in the concept of sustained yield. Rather, it was that the policies applied only to timber. Although Sloan and vir-tually every other writer of royal commission reports on this sub-ject acknowledged the necessity of caring for non-timber forest values, the policies and practices adopted by provincial govern-ments after the war focused almost exclusively on trees and tim-ber production. The sustaining of other forest values – fish and wildlife, water quality, and biological diversity – was not part of the agenda.

In spite of their flaws, the policies adopted by the provinces after World War II did accomplish one overwhelmingly impor-tant objective: they brought to an end the unrestrained harvesting of the country's forests and the neglect of forest land that had been logged, burned, or destroyed by insects. The reduction of the forest land base was reversed, as was the decline of the timber inventory.

The maintenance and care of the country's permanent forest became a universally accepted priority. Whether the harvest levels determined under these policies were the correct ones was a different matter, and one that could be rectified at a later date. The great achievement was adoption of a concept, a methodology, and above all, a desire to maintain forests in perpetuity.

SUSTAINABLE FOREST MANAGEMENT

By the late 1970s the shortcomings of Canada's sustained yield policies were increasingly apparent. A series of crises had developed, the very nature of which were often a matter of intense disagreement.

In some parts of the country the consequences of past forest practices created problems. The selective high grading of pine from Newfoundland, Nova Scotia, and New Brunswick for more than two hundred years had produced forests composed of less-valuable species that were, as they matured, highly susceptible to insect attacks. Massive infestations in Newfoundland, Quebec, and New Brunswick had been fought with little success by the aerial spraying of insecticides. By this time the warnings against widespread use of insecticides such as DDT (dichlorodiphenyltrichloroethane) in Rachel Carson's 1962 book *Silent Spring* were in wide circulation. When a similar spraying program was proposed in 1976 to combat a spruce budworm outbreak at Cape Breton, Nova Scotia, a public protest against the use of insecticides began that eventually spread throughout the region and broadened its scope to include other forest-practice issues.

For the fibre-based forest industries, the looming predicament was perceived as one of timber supply. In some cases the uncertainty about future supplies, created in many cases by the inadequacy of provincial forest inventories, threatened the decades-long expansionist mode of the industry. To many in industry, government, and communities dependent on the industry, steady increases in the production of lumber, plywood, pulp and paper had become the norm. Stability had become confused with continuous expansion. In other regions the supply problem was more immediate, as mill closures raised doubts about the ability of the established industry to maintain existing operations.

At this time, steady growth in global demand for forest products was projected, with timber shortages anticipated by the year 2000. Many in the Canadian forest industry saw this growing demand as an opportunity, and any possibility of timber supplies not being adequate to meet it as a crisis. That the projected increase in demand did not fully materialize was irrelevant; in the late 1970s it was perceived as an opportunity not to be foregone.

Nationally, there were several components to the supply problem that had developed under sustained yield. At the most basic level, the difficulty lay with a simple lack of forest-inventory information. The provincial governments that owned the forests were unwilling to provide the funds to undertake the required inventories. The extent of some forests, their current condition, rate of growth, and so on was not known. Therefore, it was difficult or impossible to know if harvest levels were sustainable or not.

Other problems lay with the concept of sustained yield itself – or the version of it adopted in some provinces. The initial objective of regulating annual harvests was not to establish an allowable annual cut that could be maintained in perpetuity. Rather, it was to harvest the existing natural forest in such a way as to create sustained-yield forest units with an even distribution of age classes. In theory, a portion of the forest unit would mature each year, be harvested, and promptly reforested. The result at some point in the future would be a steady, perpetual yield of timber that would be the optimum volume of timber the unit was capable of producing.

By the 1970s, however, it was apparent that the long-term sustainable yield of these units would be less than the volume obtainable from them through the initial harvest of old-growth timber. This decline in timber volume is known as the "fall down." Although no one doubted the validity of this concept, the rate at which existing harvest levels should be lowered to the long-run sustainable yield was debated vigorously. Some argued that cut levels should be lowered to sustainable levels immediately, others that they should be lowered over a period of decades, and others that they should not be lowered until all old-growth timber had been cut. In any case, by the 1970s, in many forests it was possible to calculate the extent and timing of the fall down.

A similar debate ensued over cutting rates in some of the

country's older forests, such as the moist cedar-hemlock forests in the Pacific coastal region where fires rarely occur. In other forests, where fires had been fought successfully for several decades, there were large areas of "unnaturally" old trees which, without protection, would have burned from lightning fires. The common factor in all these forests was that their net annual increase in timber volumes had slowed, ceased, or become a negative value. In terms of timber value they were "stagnant" or "over-mature" forests. Many foresters believed they should be harvested as quickly as possible, even at rates above long-run sustainable levels, and replaced with healthy, vigorous young trees.

A major concern across the country, especially in Crown-owned forests, was the amount of unstocked land. Whether the forest cover had been removed by fire, insects, disease, or logging, there was a vast and growing area that had not been reforested with commercially useful tree species. A report on the country's timber supply prepared for the 1980 Forest Congress said variations in reporting methods among provinces made it difficult, if not impossible, to calculate the amount of "not sufficiently restocked" forest land created during the previous decade, not to mention that accumulated in previous decades. Additionally, the report stressed, corrections in allowable annual cuts needed to be made for withdrawal of forest lands from commercial use for parks and wilderness.

The report concluded: "For Canada as a whole, the combined impact of the recent reductions in the timber base would appear to require a downward revision (of allowable annual cuts) of 15 per cent. Some of these have already been made … and the forest sector can have an assured future supply of timber, provided that obvious conditions are met. First, there must be adequate forest renewal and forest protection programs. Second, any further withdrawals of forest land must be held to a minimum. That is the forestry challenge of the 80's."

Historically, the response of the timber industries to this sort of supply crisis was to go further afield for logs. In British Columbia, loggers were dispatched into the more remote areas, to the heads of mountain valleys and into the undeveloped regions in the north of the province. Elsewhere, roads were pushed north into the boreal forests from Yukon all the way to Labrador. The economics of this long-standing strategy were problematic. Timber costs increased, forcing operational closures earlier in the cycle of declining markets.

A more serious problem was that, as it moved into previously untouched forest regions, the industry increasingly came into conflict with other forest users. On the B.C. coast there were growing conflicts with fisheries interests anxious to protect the spawning habitat of anadromous fish. In the northern forests conflicts multiplied with aboriginal people who depend on forests for food supplies and furs for their cash income.

The most serious conflicts, however, were with the rapidly growing population of affluent urban residents who used remote forests for recreational activities. These people, in combination with a science-based environmental movement, created a powerful protectionist lobby that harked back to the days of John Muir. Across the country, and especially in British Columbia, a series of battles over specific tracts of forest erupted. The forest industry, and provincial governments that had granted leases to harvest timber on Crown land, were confronted with demands for a ban on industrial activity in these areas and their designation as parks or wilderness preserves.

As advised in a report on the 1980 Forest Congress, industry resisted these demands, attempting to minimize the creation of new parks and other protected areas. But political pressure mounted, and as governments agreed to these demands, an intense conflict between the forest industry and environmentalists erupted. With each successive loss of industrial forest land, the timber supply declined further.

During the 1960s an operational principal known as "multiple use" had evolved to determine how forests could be used by different interests. In theory it allowed for a variety of uses and values – recreation, wildlife, fisheries, watershed protection, timber, grazing, and traditional cultural uses. By the 1970s, however, it was apparent that in practice multiple use did not work. Some uses reduced other values or precluded other uses, at least in the short run. Timber harvesting, for instance, could reduce or destroy the recreational value of a forest for several decades;

creation of a park excluded hunting. Various planning processes were devised in an attempt to integrate the management of different values or uses, but increasingly this proved impossible. As Bill Young, B.C.'s chief forester, told the 1980 Forest Congress: "Advocates across the nation are expecting more emphasis on wilderness areas, parkland, agricultural expansion, wildlife protection, water protection, aesthetic values, and, yes, even demanding more wood to be available for harvesting. In an increasingly greater part of the nation, the forest land base is simply not capable of meeting all of these demands." Finally, after several centuries of development, Canadian demand had reached or, in some cases, exceeded the limits of forest use.

The responses to this situation were varied. Provincial governments dramatically increased their reforestation budgets. The federal government ended its program of funding provincial construction of logging roads into increasingly remote areas and entered into joint programs to fund intensive forest management activities. A series of Forest Resource Development Agreements (FRDA) were concluded with the goal of increasing the timber supply through increased levels of silviculture. Unstocked forest lands were reforested, and other silvicultural treatments such as the spacing and fertilization of trees, intended to increase growth rates, were implemented.

Although several hundred million dollars were expended on FRDA programs, they were of limited success. The major accomplishment was the reforestation of a portion of the backlog of unstocked or poorly stocked forest lands and, incidentally, the establishment of a strong reforestation ethic in the country. Although replanting of logged and burned lands increased dramatically after 1970, the area of inadequately stocked forest land continued to increase – in part because of a definitional broadening of the term.

Some of the silvicultural programs funded under FRDA were not nearly as successful, primarily because the country lacked the silvicultural capability required to implement them. Skilled workers did not exist, centralized government management agencies were incapable of efficiently administering operational level silviculture, and the tenure system on Crown forests provided insuffi-

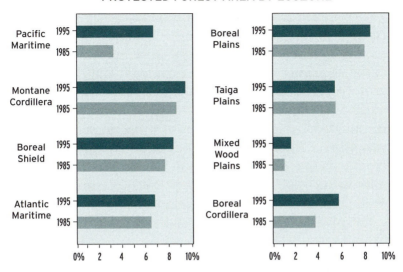

PROTECTED FOREST AREA BY ECOZONE

Figure 5.1 The amount of protected forest increased in most forest regions during the decade 1985 to 1995. Source: Environment Canada

cient incentives for the private sector to invest in silviculture. The last of the government-funded intensive management programs were terminated in the mid-1990s.

The partnership between industry and government, designed decades earlier to facilitate timber harvesting on Crown forest land, proved to be unsuitable in most cases for long-term management of forests. Governments that owned the forests were unwilling to maintain the long-term financial commitments required, and industry was unwilling to invest in forests it did not own without guarantees that it would receive the benefits of those investments. Governments, however, increased their forest revenues throughout this period. Corporate income taxes paid to the federal government increased seven-fold between 1970 and 1987, and taxes paid to provincial governments increased more than five-fold during the same period.

In response to public and scientific pressure throughout the 1980s and especially the 1990s, many provinces gave protected status of one sort or another to increasing amounts of forest. Parks, wilderness areas, and ecological reserves were increased in number and size. Another category of protection was afforded by operational policies applied to industrial land outside protected areas. This included forest land in riparian areas, wildlife winter habitat, bird-nesting sites, steep slopes, rocky outcrops, and other small areas within the working forest where special, more costly harvesting practices were required. By the end of the century these areas comprised one-fifth of the country's productive forest land.

Beginning in the 1970s governments, voluntarily or in response to growing demands, provided opportunities for public participation in forest planning processes. Typically, following the creation of forest reserves early in the twentieth century, decisions on forest use had been made in isolation – or in consultation with the forest industries – by provincial forest services.

After World War II, as the frequency of conflicts between timber industries and other forest users increased, provincial forest services began consulting other public resource agencies, while retaining the final decision-making authority. Thus, a provincial forest service might consult the federal fisheries department before approving a permit to log near a salmon spawning stream.

By the early 1970s some provinces had developed a planning process that sought to integrate management of the various resources and values present in the forest area under consideration. This process was usually open only to the various resource agencies concerned with these values, and public-interest groups were expected to work through these agencies.

This process could not accommodate the enormous increase in public interest in forests that began to develop at this time. Governments were forced to establish processes in which everyone could participate. This placed enormous pressures on provincial agencies, especially forest services, to modify their operational procedures. In many cases this led to ruptures in the longstanding working relationships between the government agencies and the timber industries. Governance of Canada's forests entered a period of confusion and conflict.

Opening up the planning process to include a wider range of interests in discussions about forest use ultimately produced new perspectives on forests and how they should be managed. The focus of attention on specific species or classes of species, for instance, previously manifested through organizations and government agencies concerned with game fish or wildlife, broadened. Now the focus was on forest species in general and the forest habitat they require to survive. The emphasis shifted from maintaining populations of certain species to maintaining the diversity of all species found in a given forest. By the early 1990s this new way of looking at forests, as habitat for the entire range of plant and animal species it hosts, including timber species, created a concept of how to use forests reminiscent of the ideas of Aldo Leopold.

Known variously as "holistic" or "wholistic" forestry, "ecoforestry," and "sustainable forest management," this concept is a logical extension of the conservationist concept of the permanent forest and a broadening of the concept of sustained yield to include all the components of a forest. As Herb Hammond, a B.C. forester, wrote in 1991: "In the wholistic view, we acknowledge the importance of all life forms and we fully protect biological diversity in order to maintain healthy forests and healthy communities. This approach ensures that we will not foolishly squander the forests, but instead will protect the forests for now and for future generations."

A decade later Ray Travers, a Victoria forester and former chair of the Ecoforestry Institute, wrote in a similar vein: "Ecoforestry is forestry in nature's image, where nature is not just a set of limits but is a model for the design of practices which maintain the health, diversity and productivity of the forest, upon which all life and human prosperity depend. The core values of Ecoforestry are ecological integrity, community vitality and economic opportunity."

And, from a broader perspective, the Canada Forest Accord, a commitment to sustainable forest management drafted in 1998 and signed by a broad coalition of provincial governments, timber-based forest industries, academics, professionals, and other forest users, stated: "Our goal is to maintain and enhance the long-term health of our forest ecosystems, for the benefit of all living things both nationally and globally, while providing environmental, economic, social and cultural opportunities for the benefit of present and future generations." This is the goal of the National Forest Strategy adopted at the 1992 Forest Congress and refined in 1998 and 2003. The strategy is more than a mere state-

Figure 5.2 Variable retention logging attempts to maintain the biological diversity of forests by leaving portions of the original forest throughout the site. Source: Vancouver City Archives AM 1386, 599-F-3

ment of principles and goals; it includes commitments to action. These include the establishment of a network of eleven model forests in the country's various forest regions to serve as working models of sustainable forest management. Several provinces have framed codes that define the legal requirements of sustainable forest management. In 1995 a set of national "criteria and indicators" was devised to define the concept of sustainable forest management, guide its implementation, and measure progress. An independent panel of experts appointed to review the national strategy's progress concluded in 1997 that "Canada probably is moving toward sustainable forest management, but at the moment this is only measurable in terms of local forests, and not yet on a national scale."

By the end of the century that movement was discernible in a number of on-the-ground developments. Several forest companies began the move away from clearcut logging and adopted a silvicultural system called "variable retention harvesting." Its primary objective is to maintain the biological diversity of forest sites, and it does so by leaving intact portions of the original forest – individual trees, groups and clusters – to provide habitat for the various species occupying the site.

Certification of products is another initiative that evolved out of the interactions of forest companies and activists, when the latter initiated campaigns abroad to boycott Canadian forest products. It entails a process of auditing forest practices and certifying that the products offered for sale have been manufactured from timber obtained from well-managed forests. Certification provides individual companies with incentives to provide high levels of forest management. By the end of the twentieth century, most Canadian forest companies had, or were in the process of having, their operations certified by one or more of the national and international certification schemes available.

Figure 5.3 Designed for selective logging of steep slopes, this Timbco harvester is typical of new forest technology built to sustain the health of forests. Source: author's collection

The idea of sustainable forest management is at best an evolving concept. It is not a fixed set of rules or principles that apply to all forests, for all time. In practical terms, sustainability means different things in different forests, or even in different parts of the same forest. And because forests are dynamic entities, what constitutes sustainability in a given forest today will not necessarily hold true a decade or a century from now. Sustainability is more a state of mind or an attitude than a set of rules and regulations. As Peter Murphy, a University of Alberta forestry professor, wrote in 1998, "Sustainable forest management is not so much a destination as it is a journey."

State of the Forests: 2000

After ten thousand years of human use, Canada's forests still exist virtually intact. Beginning as a few isolated pockets of tree-covered land, they advanced in the wake of receding glaciers to cover much of the Canadian landscape. Since their creation, they have been subject to human manipulation, so that it is impossible to know with any great degree of accuracy the extent of the forests when European colonization began about five hundred years ago. Rough estimates have been made based on calculating the full area of each forest region, but these do not take into account areas cleared by aboriginal people.

There is also a definitional problem of deciding what constitutes forested land. In the north and south central regions, an ill-defined border divides the boreal forest from the open tundra and grasslands. Current conventions define forests as land with crown cover of more than 10 per cent by trees more than seven metres tall. Earlier estimates used different definitions, so it is difficult, if not impossible, to compare current calculations of forest area with those of other times.

With these definitions and cautions in mind, it has been estimated by the World Resources Institute that since European colonization of Canada began, 26.5 million hectares of forest – about 6 per cent – have been converted to other uses, including agriculture, urban land, water reservoirs, transportation and utility corridors, and mine sites. Today there are almost 418 million hectares of forest, indicating a pre-colonial forest of about 444 million hectares.

These conversions have been unevenly distributed between forest regions and between provinces. The largest areas of permanently deforested land are found in Ontario and the three prairie provinces, especially Alberta and Saskatchewan. These constitute

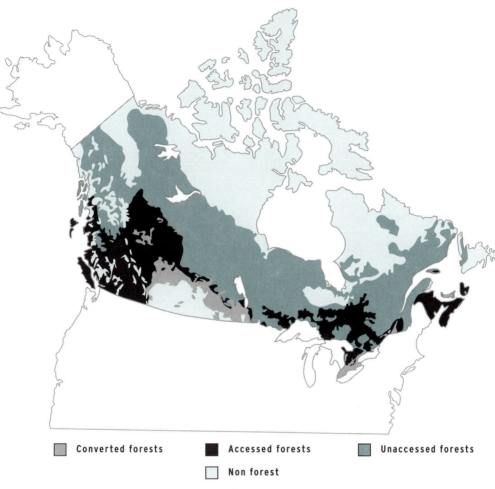

Figure 5.4 The conversion of Canadian forests to other uses, and increased accessibility to the remaining forests, has occurred unevenly across the country. Source: "Canada's Forests at a Crossroads"

the heavily developed agricultural areas in the Carolinian and Great Lakes forests, and the aspen parklands that lay between the central boreal forest and the prairie grasslands. At the same time, it should be noted that the eastern provinces have more forested land than seventy-five years ago, largely due to the reforestation of abandoned agricultural lands. Only 3 per cent each of the coastal, Columbian, and boreal forests have been converted to other uses. None of the northern boreal or the subalpine forests has been converted.

Forest Conversion by Province, 2000

Province	Converted land (000 hectares)	Percentage converted	Total remaining forest (000 hectares)
Northwest Territories	1	0	46,517
Yukon Territory	1	0	12,929
British Columbia	849	2	48,184
Quebec	1,468	2	80,948
Newfoundland	15	0	16,813
Ontario	3,858	5	78,721
Nova Scotia	203	4	4,879
New Brunswick	239	3	6,612
Prince Edward Island	211	46	250
Alberta	9,035	21	33,259
Saskatchewan	7,270	19	31,558
Manitoba	3,304	8	40,289
Nunavut	0	0	1,977
Total	26,454	6	402,936

Figure 5.5 Conversion of forests to other uses, primarily agriculture, has been concentrated in some areas, totalling only 6 per cent of the original forest. Source: "Canada's Forests at a Crossroads"

The fact that Canada has retained 94 per cent of its forest land makes it unique among major forest nations. The United States has retained about 70 per cent of the forests it had in 1600, Russia almost 69 per cent, Finland 82 per cent and Sweden 86 per cent. A significant difference between Canada and these countries is that Canada has a much higher proportion of large, undisturbed forest ecosystems, most of them in the northern boreal forests.

Another encouraging sign is that, according to the World Resource Institute, Canada's forest coverage has increased slightly during the past decade, at an annual rate of .07 per cent. Most of this new forest area is in marginal agricultural regions where land has been planted to trees or has afforested naturally. It does not include a large and growing area of urban forests established and maintained in towns and cities across the country.

Canada has one of the largest forests, in relation to its population, of any significantly forested country in the world. In 2000 there were about 13 hectares of forested land for each of the

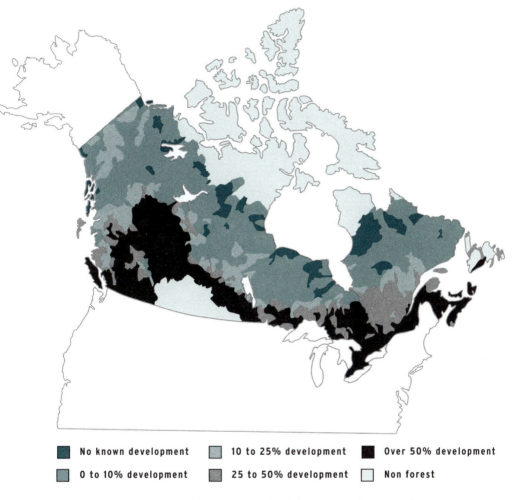

Figure 5.6 Development of forested watersheds has occurred to some degree in most of Canada's forests. Source: "Canada's Forests at a Crossroads"

country's 31.4 million people.

Almost two hundred active mines are located within Canada's forests, eighty of them in the boreal and thirty-eight in the Great Lakes forest. There are eleven hundred hydroelectric dams, about two-thirds of them in the same regions. Nationally, more than four thousand settlements are located in forested areas. Overall, 51 per cent of the country's forests are situated within ten kilometres of a development activity.

Forest fragmentation in Canada, although low compared to

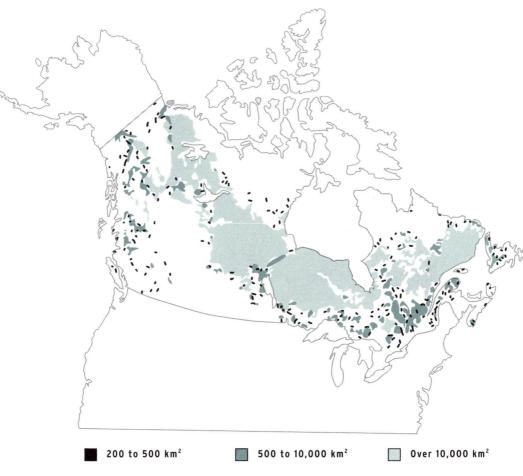

| ■ 200 to 500 km² | ▨ 500 to 10,000 km² | □ Over 10,000 km² |

Figure 5.7 The large remaining areas of unfragmented forest are found mostly in the boreal forests. Source: "Canada's Forests at a Crossroads"

other countries, continues at an accelerating pace. The construction of roads, pipelines, railways, power transmission lines, and recreational access routes has a variety of impacts on the forests. They become divided and opened up to a wide range of activities previously excluded because of inaccessibility. The highest levels of fragmentation are in the southern and coastal forests where development has been most intensive. A distinctive type of fragmentation, perhaps the heaviest outside urban areas, has occurred in northern Alberta and northeastern British Columbia through construction of petroleum exploration lines, well sites, and access roads – almost all in the past fifty years.

The most concentrated areas of unfragmented forest are in the central and northern boreal, the northern interior of B.C., and the Great Lakes forest regions of Ontario and Quebec. Today, more than 60 per cent of the country's forests are in large, unfragmented tracts of ten thousand square kilometres or larger. Less than one-seventh is in tracts smaller than two hundred square kilometres.

Forest-fire policies and practices have undergone continuous evolution. Beginning in aboriginal times, extensive burning of forests was tolerated, even encouraged. The adoption of conservation policies in the early 1900s included a campaign to convince aboriginal people, settlers, railway operators, and others that fires were wasteful and should be prevented and, if possible, extinguished. By the end of World War II, this policy was firmly established, and the technology was available to protect forests from fire. Fire forecasting techniques were refined and widespread use of aerial reconnaissance and suppression systems developed. For the past fifty years the capability to detect fires in their early stages and extinguish them has increased enormously, although fires still consume far more timber than the forest industry. Over the past three decades the annual number of fires has remained fairly constant at an average of about nine thousand a year, burning on average two million hectares a year. In

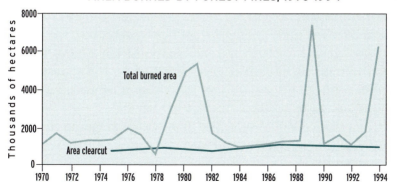

Figure 5.8 Since 1970, the area clearcut-logged has remained relatively constant. About 80 per cent of the much larger burned area consists of commercially non-productive forests. Source: "Canada's Forests at a Crossroads"

some years – most notably the early 1980s, 1989, and 1995 – the areas burned increased dramatically.

The recent success at fighting fires has created conditions in some forests that have thrown into question fire policies of the past century. In the eastern boreal, fire suppression, along with harvesting practices, has increased the balsam fir content of forests, leading to increased attacks on the species by budworms; this, in turn, increases the danger of large, catastrophic fires. A similar situation has evolved in the montane forests of the B.C. interior where increased fire-suppression capability has led to a build-up of older lodgepole pine forests that are susceptible to mountain pine beetles. In 1998 a beetle infestation began in and around Tweedsmuir Park. Five years later it had heavily infected almost nine million hectares in west-central B.C., creating ideal conditions for a catastrophic wildfire. These and similar situations elsewhere in North America have led to a basic re-examination of fire policy in Canada, signifying a move towards the purposeful use of fire and away from policies stressing its elimination from the forest.

During the past decade, partially in response to international initiatives and protocols, Canada has agreed to a series of measures that will protect certain defined areas including forest lands from development. Because these agreements are made by the federal government, and the authority to create parks and other protected areas is in provincial jurisdiction, this has been a slow and difficult process. By 2000 about 32 million hectares – 7.7 per cent of the country's forest – had been given protected status. Governments have been less nimble in providing funds for the care of these areas. Policies governing the care and management of the country's protected forests are poorly developed and a matter of growing concern.

In 1992 federal and provincial governments agreed to create an ecologically representative network of protected areas by the year 2000. There are 388 forested natural regions in the country, and by 2000, 60 per cent had been partially or fully represented, leaving 40 per cent unrepresented within protected areas.

An estimated 125,000 species of plants, animals and microorganisms occupy – in fact, constitute – Canada's forests. Of these, according to the World Resources Institute, eighty-four are considered to be at one degree of risk or another. Since cataloguing of species began, two forest species – the passenger pigeon and the Queen Charlotte Islands woodland caribou – have become extinct. Another four species – one plant and three butterflies – are extirpated (no longer exist in the wild). According to the Committee on the Status of Endangered Wildlife in Canada (COSEWIC), of the at-risk species that live in the forests, twenty-two are endangered (facing imminent extirpation or extinction), twenty are threatened (likely to become endangered if limiting factors are not reversed), and forty-two are vulnerable (sensitive to human activities or natural events). These low numbers, especially of extinct forest species, are generally not considered indicative of the risks faced by forest species. Local and regional extinctions of many species have occurred – virtually all large forest predators in Prince Edward Island, for instance. Continued forest disturbance increases the risks to many local populations.

Canadian Forest-Dwelling Species at Risk, 1999

Categories	Mammals	Birds	Plants	Reptiles	Total
Endangered	3	4	14	1	22
Threatened	2	4	11	3	20
Vulnerable	13	9	14	6	42
Total	18	17	39	10	84

Figure 5.9 Source: Canadian Forest Service

The primary focus of the new concept of sustainable forest management is the maintenance of species diversity everywhere. Every extinct and extirpated forest species and most at any kind of risk attained that status under management policies that made little attempt to sustain non-timber values.

Industrial forestry is permitted in about 56 per cent of the country's forests. Although 245 million hectares are classified as timber productive and suitable for commercial forestry operations, a growing portion of this area – 10 million hectares in 1991 – is not available for harvesting because of environmental constraints. About 42 per cent of the Canadian forest is considered mature or overmature and 46 per cent immature.

The area harvested each year increased steadily from less than 400,000 hectares in 1920 to an all-time high of 1.1 million hectares in 1988, after which it stabilized at about 1 million hectares. Since the industry began in the early seventeenth century, about 45 million hectares, or less than 20 per cent, of the commercial forests have been logged, some of them more than once.

Although there is no estimate of the timber volumes found in Canada's forests when colonization began, according to Canada's Forest Inventory there is now an estimated 26.2 billion cubic metres of mature timber in the country's timber-productive forests, 77 per cent of which is in coniferous species and 23 per cent in broadleaved species.

Current growth rates in timber-productive forests average 1.59 cubic metres of merchantable timber per hectare a year, or a total of 360 million cubic metres annually. This volume is almost double the rate of harvest, indicating a steady increase in the country's standing timber inventory. This annual increase, however, fluctuates dramatically due to the destruction of timber by fire, insects, and disease. Timber growth in almost all forests is unassisted, and with increased levels of silviculture could be increased substantially to maintain or enhance the timber supply.

One type of forest management in which Canada has excelled is reforestation after harvesting or fires, with substantial improvements made over the past twenty-five years. The area replanted or seeded increased from 150,000 hectares in 1975 to almost 500,000 hectares in 1995 – a much more rapid increase than the area logged annually. Similarly, the amount of stand tending increased eight-fold during this period, although the actual amount is still small and the benefits obtained are unclear.

The yield of mature timber from the country's forests varies dramatically, ranging from an average of 95 cubic metres a hectare harvested in Nova Scotia to as much as 700 cubic metres per hectare in B.C.'s coastal forests. In 1997, 182.7 million cubic metres of timber were harvested from 1.02 million hectares – an average of 179 cubic metres per hectare. This is considerably lower than the allowable annual cut for that year of 236.5 million cubic metres.

What timber harvest levels might be after widespread implementation of sustainable forest management is unknown. Current allowable cut calculations do not take into account potential changes in timber yields due to the adoption of new management practices. Some analysts estimate that the adoption of sustainable forest management practices will require harvesting constraints that will reduce timber harvests by as much as 50 per cent, while others suggest these practices can double or triple timber growth rates and harvests, as well as improving timber quality. Experience elsewhere suggests that timber yields can be increased many times over normal yields. Whether this is possible while maintaining non-timber values is unknown.

Of equal importance in considering the future of the country's forests is the question of global supply and demand for forest products, matters beyond the control or influence of Canadian policy-makers and the forest sector. At the beginning of the twenty-first century, a long-term timber surplus appears to be building. Large volumes of industrial timber from plantations in the southern hemisphere are becoming available, as are supplies from the southeastern United States, Europe, and elsewhere. Although global demand for timber is increasing, so is the supply, and the world demand for Canadian forest products may not be as great as in the past.

The costs of maintaining intact natural forests, as indicated in the concept of sustainable forest management, will likely be higher than under previous management systems and probably much higher than those incurred in plantation forests abroad. Forest managers will need to be extremely efficient, innovative, and dedicated to compete with timber producers in other countries who operate under less demanding conditions. Large, ongoing investments in forest management will be required. To date, the provincial governments that own the forests have not shown a willingness to make such investments and to maintain them over the long periods of time involved. Nor are there incentives for private investment in such management. As the authors of a 2000 paper on the future of Canada's forests, *The Perpetual Forest*, state, "Crown tenure policies have, generally, failed to provide the incentives needed for the private sector to invest in the types of forest management needed to enhance the timber supply . . . The policy framework has been even less capable of providing incentives for the kinds of forest management needed to protect and enhance the full range of forest values found across the country. The public has lost confidence in the type of forest administration possible under these policies; and some of these policies are at the centre of an ongoing dispute with our largest forest products customer. Clearly, it is time they were reformed."

This policy reformation has barely begun, and carrying it out will require sustained effort on the part of all Canadians. Long-term investments in forest management must be made to avoid reductions in timber supply as well as non-timber values. During the past decade, provincial governments have faced escalating and competing costs that have hampered their ability to make such long-term investments. Caring for forests that do not produce revenue – protecting them from fires, insects, disease, and illegal use – inposes a further financial burden on cash-strapped governments at a time when reductions in funding for the upkeep of parks and other protected areas have already become commonplace. Renewed commitments toward stewarding Canada's forests and recognizing their value are necessary, and the plans for purposeful management described in Canada's National Forest Strategy are a clear indication that Canadians are willing to assume such responsibility.

As has been the case for the past ten thousand years, Canada's forests are in a state of constant change. Much of the change occurring over the past five hundred years has been due to human activities, at rates that have accelerated until recent times. It is undeniable that these changes, including the 6 percent conversion of forests to non-forest status, have occurred. By the same token, it is inaccurate and misleading to claim that the country's forests have been, or are in the process of being, destroyed, devastated, decimated, or damaged beyond recovery. For the most part – perhaps here more than any other place in the world – they are whole and healthy.

Most of the damage inflicted on the country's forests occurred during periods when public opinion and forest policies were based on values and priorities far different from those that prevail at present. Most of this loss took place when forests were not

Figure 5.10 In the 1860s the hills around Barkerville in central B.C. were denuded by gold miners. Since then the forests have undergone more than a century of undisturbed regrowth. Sources: B.C. Archives 10070 and Vancouver City Archives AM 1386, 599-E-3

valued or when they were seen as obstacles to the development of the country. During this time two species have been lost forever – a lamentable but understandable occurrence, given the circumstances of the day.

Today a new attitude and a new set of forest policies derived from it prevail, which seek to sustain the forests and all the species that consitute them. The question now facing Canadians is not *whether* to use a particular forest but *how* to use it in a manner that will sustain the full range of its productive diversity in perpetuity.

CONCLUSION

The guiding principles of forest use in Canada and the practices derived from them have gone through many changes over the past century and continue to evolve. In 1900 the prevalent attitude was still that of liquidation, with only a minority concerned with forest conservation. By mid-century, conservationist thinking was well entrenched, but the idea of limits, voluntarily accepted before the forest resource was utilized beyond redemption, was reluctantly adhered to. In time the concept of sustained yield was adopted, only to be revealed as wanting in some respects.

The acceptance of dramatic new ideas by democratic societies such as Canada is a slow process. Putting these ideas into practice, once they have been accepted, can take even longer. As H.R. MacMillan noted in the 1930s, conservationist principles had not been converted into widespread practice fifty years after they had become widely understood. Even today, sustained yield, while accepted in theory, has yet to be fully realized, as the inability to maintain harvest levels in some parts of the country attests.

The idea of sustainable forest management is not new. Half a century ago Aldo Leopold proposed essentially the same concept. It lay more or less dormant until about twenty years ago. Now it has become part of mainstream thought, and throughout the country thousands of people who work in forests are attempting to translate it into practice.

This is not an easy task. It is difficult enough for people to change the way they think about the work they do. It is also difficult to reform the forest economy without inflicting hardship on individuals and communities. It is even harder to revise the laws, regulations, practices, industrial structures, and social conventions that will need changing before sustainable forest management begins to achieve the goal set out for it in the Canada Forest Accord.

In choosing to manage their forests in the manner outlined in this strategy, Canadians have set themselves an immense challenge. Because the revenues required to fund it must be earned in a global marketplace that includes producers who are not similarly constrained, Canadians will be faced with obtaining an industrial timber supply from natural fully diverse forests in competition with producers utilizing cheaply grown plantation timber or, in some cases, timber drawn from natural forests utilized in an unsustainable manner. They will be the first to learn the strengths of the concept of sustainable forest management, and the first to see its weaknesses. Their success will depend largely upon the understanding and support of all Canadians who, if the objective of sustaining the county's forests is to be met, will necessarily be participants in the challenge.

A century ago few people except those directly engaged in the forest industries felt any great concern about the country's forests. Today there is a broadly based engagement of many people – forest workers, professionals, members of labour unions, communities, people in voluntary organizations, politicians, scientists – participating in the determination of how the forests will be used. The continued attention of these diverse interests provides a large measure of confidence about the future of the forests.

As well, Canada's forest heritage is increasingly seen as a global heritage. People from many parts of the world look to the well-being of Canada's forests as essential to their welfare. Increasingly, Canadians accept their responsibility to the rest of the world to care for these forests.

From earliest times, Canada's magnificent forests have presented challenges to the people who use them. Beginning with the original aboriginal inhabitants and continuing with the first European settlers and their descendants, Canadians have met those challenges by adopting policies that reflected the needs of their times. They must continue to meet these challenges as life on the planet evolves and needs change. Markets, which reflect human needs from the forests, will change. Population levels of all species, including humans, will change, as will the global climate. All these changes will affect Canada's forests in ways we can barely imagine, let alone control.

Today, after centuries of forest use, Canadians have accepted their role as stewards of the forests that define this country. They have committed themselves to maintaining them, enhancing them, and passing them on to succeeding generations.

SOURCES

Assessing the state of a nation's forests is a difficult task under the best of circumstances. Forests are nebulous and dynamic entities. Rarely is it possible to determine precisely where forests begin or end. They are in a state of constant, but unpredictable change. The most that is possible is an approximate snapshot; and even that requires a highly developed capability to monitor, measure, compile, and analyse data that, almost by definition, is outdated immediately after it has been gathered.

The enormity of the task is evident in the periodic attempts by the Food and Agricultural Organization of the United Nations to report on annual global production and consumption of forest products, and in its reports issued every decade on the condition of the world's forests. These reports are, in the end, only a compilation of data supplied by each country or an estimate made by foreign consultants. In some cases the available information is a fairly reliable representation of that nation's forests. In others, the data is inaccurate, hopelessly out of date, or non-existent. At any given time some countries are in a state of political upheaval and unable to supply information; others have little interest or need to gather it. In the end, by the time the information is obtained, compiled, printed and circulated, all the forests of the world have changed. Some have been logged, others burned. Some have decreased in area, others increased. Some have died.

The situation is similar in attempting to make a comprehensive overview of Canadian forests. They constitute one of the largest national forests in the world but are particularly diverse in their composition. They fall under the jurisdiction of at least eleven governments, and even this condition is fluid as the territories gradually gain control of their forest resources.

The federal government has limited jurisdiction over forests and therefore no mandate to develop a comprehensive understanding of them. Since 1990 it has published an annual report, "The State of Canada's Forests," a useful and informative document but, like most such endeavours, relying on information supplied by provincial government agencies, some of it twenty-five years or more old. For this same reason, the basic question of how much forested land there is in Canada and whether it is increasing or decreasing has until recently been virtually impossible to answer. No one has had sufficient up-to-date information to address that issue decisively.

Ironically, what is perhaps the most thorough report on Canada's forests, "Canada's Forests at a Crossroads," was issued in 2000 by a U.S.-based environmental organization. Utilizing satellite imagery, it came to a conclusion that Canadian government officials suspected was true but lacked the confidence to state as fact – that Canada's forests are increasing in area.

Little work and even less writing have been devoted to evidence of pre-contact modification of Canadian forests by aboriginal people, although a great deal has been published on this subject in the United States. The only available descriptions are in the accounts of early travellers in various parts of the country, who mention in passing their observations of sizable forest clearings or provide eye-witness accounts of natives burning forested areas. The sheer size and diversity of Canada's forests have also inhibited academic and popular writers from attempting to describe them in their totality. Many books have been written on how forests have been used – or, more often, misused – in many parts of the country, and in a few cases, in the country as a whole. When it comes to descriptions of the forests themselves, most authors have restricted themselves to a particular province or region. In this manner, much of the country's forests have been covered. What is lacking , however, is an overview and, consequently, an understanding of the Canadian forest as a coherent phenomenon.

This book draws heavily on these disparate sources. It is not based on original scientific or academic research or on a vast experience in all the nation's forests. It attempts to piece together an understanding of this huge subject by stitching together bits and pieces of information about different parts of the Canadian forest recorded by many writers over the past century or more.

The sources most crucial to this examination include the following works.

Historical Data

Gentilcore, R. Louis, ed. *Historical Atlas of Canada*. Toronto: University of Toronto 1993.

Leacy, F.H., ed. *Historical Statistics of Canada*. Statistics Canada and Social Science Federation of Canada 1983.

Descriptions of Forest Regions

Johnson, Ralph S. *Forests of Nova Scotia*. Halifax: Four East Publications and Nova Scotia Department of Lands and Forests 1986.

Lambert, Richard. *Renewing Nature's Wealth*. Toronto: Ontario Department of Lands and Forests 1967.

Lawrence, R.D. *The Natural History of Canada*. Toronto: Key Porter 1988.

Potvin, Albert. *A Panorama of Canadian Forests*. Canadian Forestry Service 1975.

Rowe, J.S. *Forest Regions of Canada*. Canadian Forestry Service, Pub. no. 1300, 1972.

Young, Cameron. *The Forests of British Columbia*. Vancouver: Whitecap 1985.

Historic Condition of Canada's Forests

Leavitt, Clyde. *Forest Protection in Canada*. Commission of Conservation, Ottawa, 1912.

Whitford, H.N., and R.D. Craig. *Forests of British Columbia*. Commission of Conservation, Ottawa, 1918.

Current Condition of Forests

Apsey, Mike, Don Laishley, Vidar Nordin, and Gilbert Paille. "The Perpetual Forest, *The Forest Chronicle* 76, no.1 (2000): 29–54.

"Compendium of Canadian Forestry Statistics." Canadian Council of Forest Ministers. Annually since 1991. Also see http://nfdp.ccfm.org/framesinv_e.htm

Canadian Council of Forest Ministers. "Canada Forest Accord," in *National Forest Strategy: 1992–2003*. Ottawa: Canadian Council of Forest Ministers 2000.

"Canada's Forests at a Crossroads: An Assessment in the Year 2000." World Resources Institute, Washington, 2000. Also see www.wri.org

"The State of Canada's Forests." Natural Resources Canada, Ottawa. Annually since 1990. Also see

http://www.nrcan-mcan.gcca/cfs-scf/national/what-quoi

FURTHER READING

Apsey, Mike, Don Laishley, Vidar Nordin, and Gilbert Paille. *The Perpetual Forest: Using Lessons from the Past to Sustain Canada's Forests in the Future*. Ottawa: Canadian Institute of Forestry 2000.

Bouthillier, Luc. "Quebec: Consolidation and the Movement towards Sustainability." In *Canadian Forest Policy: Adapting to Change*, edited by Michael Howlett, 237. Toronto: University of Toronto Press 2001.

Canadian Council of Forest Ministers. *Sustainable Forests: A Canadian Commitment*. Ottawa 1998.

Defebaugh, James Elliott. *History of the Lumber Industry of America*, vol. 1, *The American Lumberman*. Chicago: n.p. 1906.

Drushka, Ken. *HR: A Biography of H.R. MacMillan*. Madeira Park, B.C.: Harbour 1995.

– *In the Bight*. Madeira Park, B.C.: Harbour 1999.

Drushka, Ken, and Bob Burt. "The Canadian Forest Service: Catalyst for the Forest Sector." *Forest History Today* (spring/fall 2001): 19–28.

Gillis, R.P., and T.R. Roach. *Lost Initiatives*. New York: Greenwood Press 1986.

Hayes, Samuel P. *Conservation and the Gospel of Efficiency*. Cambridge, Mass.: Harvard University Press 1959.

Heidenreich, C. *Huronia: A History and Geography of the Huron Indians, 1600–1650*. Toronto: McClelland & Stewart 1971.

Hirt, Paul W. *A Conspiracy of Optimism*. Lincoln: University of Nebraska Press 1994.

Lower, Arthur. *Great Britain's Woodyard: British America and the Timber Trade, 1763–1867*. Montreal: McGill University Press 1973.

– *The North American Assault on the Canadian Forest.* Toronto: Ryerson 1938.

– *Settlement and the Forest Frontier in Eastern Canada.* Toronto: Macmillan 1936.

Macoun, James M. *The Forest Wealth of Canada.* Ottawa: Department of Agriculture 1904.

McKay, Donald. *Heritage Lost: The Crisis in Canada's Forests.* Toronto: Macmillan. 1985.

Murphy, Peter J. *History of Forest and Prairie Fire Control Policy in Alberta.* Edmonton: Alberta Environment 1985.

Nelles, H.V. *The Politics of Development.* Toronto: Macmillan 1974.

Pyne, Stephen J. *Fire in America: A Cultural History of Wildland and Rural Fire.* Princeton: Princeton University Press 1982.

– *World Fire: The Culture of Fire on Earth.* New York: Henry Holt 1995.

Ross, Monique M. *Forest Management in Canada.* Calgary: Canadian Institute of Resources Law 1995.

Scott, Anthony. *Natural Resources: The Economics of Conservation.* Toronto: McClelland & Stewart 1973.

Wynn, Graeme. *Timber Colony: A Historical Geography of Early Nineteenth Century New Brunswick.* Toronto: University of Toronto 1981.

INDEX